After Radical Land Reforms
Restructuring agricultural cooperatives in Zimbabwe and Japan

Rangarirai Gavin Muchetu

Langaa Research & Publishing CIG
Mankon, Bamenda

Publisher
Langaa RPCIG
Langaa Research & Publishing Common Initiative Group
P.O. Box 902 Mankon
Bamenda
North West Region
Cameroon
Langaagrp@gmail.com
www.langaa-rpcig.net

Distributed in and outside N. America by African Books Collective
orders@africanbookscollective.com
www.africanbookscollective.com

ISBN-10: 9956-551-91-0

ISBN-13: 978-9956-551-91-0

Dedication

In loving memory of Kenan Dickson Muchetu

To Shona-Pearl & Zivanai Taye Muchetu,

Acknowledgements

I want to first appreciate the support and guidance I received from Professor Yoichi Mine. You provided intellectual advice and support that helped shape not only the book but me as a person. Secondly, I would like to thank my wife, Primrose Mero, whose sacrifices and loving support throughout the development of the book is immeasurable. I am incredibly grateful for your unpaid family labour in the data entering and cleaning process. Special thank you to Walter Chambati and the Sam Moyo African Institute for Agrarian Studies (SMAIAS) for the various contributions extended towards the completion of the book. I want to further extend gratitude to my parents Mr Rogermoore and Mrs Dorcus Muchetu for their never-ending support and love (*Ndinotenda ana Nyandoro, Zvaitwa Nyamasvisva*).

It is tough to find a research project such as this book that belong solely to the author. This book is a combination of direct and indirect work by a group of several people, including friends, colleagues, and staff from the Graduate School of Global Studies at Doshisha University. Thank you for the various inputs from Stephanie Lorain, Aya Koso, Kevin Nyafwa, Ming-Ru Li, Shamiso Marange, Wonder Magoso, David Muroni, Freedom Mazwi, Ahmed Marwa, Asmao Diallo, Steve Mbcri, Kundai Dube, Rungano Muchetu, Mr Mazvinyingwa, Mrs Maziva (CACU chairperson) and many others whom I cannot mention by name. I am also forever grateful to the participants and interviewees who agreed to let a foreigner into their lives in Japanese villages of Nose Farm, Ryuo Green Ohmi and Sanbu. Special mention also goes to the CA and A1 farmers in Goromonzi district who took time to answer my long questions.

This book was only possible through the generous funding of the people of Japan through the Japanese Government Scholarship programme (*Monbukagakusho*: MEXT) and research grants from Doshisha University. Last respect goes to the Langaa Research and Publishing Common Initiative Group (Langaa RPCIG) for agreeing to publish this undertaking. I am forever indebted.

Table of Contents

List of Illustrations xiii
List of Acronyms.. xix

Chapter One
Introduction to Rural Zimbabwe
and its challenges 1

Chapter Two
Some Notes on the Theory of Peasant Cooperative 23

Chapter Three
State, Markets, and the Japanese
Agricultural Cooperatives... 69

Chapter Four
State, Markets, and the Zimbabwe
Agricultural Cooperatives... 111

Chapter Five
Field evidence; Current trends and patterns in
Japanese Agricultural Cooperative System 153

Chapter Six
Field evidence; Current Trends and Patterns
in the Zimbabwe Cooperative Movement....................... 205

Chapter Seven
Peasant Differentiation and Its Effects
on Social Economic Production 251

Chapter Eight
Restructuring agricultural cooperatives:
A New Cooperative Model for Zimbabwe 313

Bibliography 345

Index.. 365

List of Illustrations

Maps
Map 1: Map of Japan showing the
three research sites xxiii
Map 2: Research sites in Japan.................................... xxv
Map 3: Map of the study site 4:
Goromonzi district, Mash-east province, Zimbabwe xxiv

Tables
Table 1.1: Some background statistics
for the research sites 21
Table 2.1: Classification of the
Chayanovian peasantry 62
Table 2.2: Summary of theoretical frameworks........................ 66
Table 4.1: Source, type and effectiveness
of land demand 1997-2020.. 116
Table 4.2: Growth of agricultural service
and collective cooperatives 1956-1987............................... 121
Table 4.3: Contemporary local level
cooperative challenges in Zimbabwe.................................... 148
Table 5.1: List of interview participants in Japan.................... 154
Table 6.1: List of structured interview
participants in Zimbabwe.. 207
Table 6.2: Attitudes towards state
hegemony in the cooperative movement................................. 239
Table 7.1: Variables used to determine
critical study components (N=192)..................................... 253
Table 7.2: Gender and marital status
of cooperative members 259
Table 7.3: Employment status of members 260
Table 7.4: What was the main reason
for joining Cooperative?... 263
Table 7.5: Founder of the cooperatives 263
Table 7.6: Determinants of someone
joining a cooperative? 264
Table 7.7: Access to information within

the cooperative.. 267

Table 7.8: Management's handling of
corruption cases... 271

Table 7.9: Cooperative debt recovery mechanisms 272

Table 7.10: Running of cooperatives
and auditing of accounting books. 273

Table 7.11: Concerns raised by co-operators
with regards to the efficiency of managers 274

Table 7.12: The strength of relations between
cooperative and other sub-sectors 275

Table 7.13: Sources of price information
for the cooperative members.. 277

Table 7.14: Access to government public
extension services. 278

Table 7.15: Use and frequency of public
extension services by cooperative members............................. 279

Table 7.16: Ease of accessing credit from
the cooperative structures 281

Table 7.17: Presence and nature of challenges
in accessing credit for agricultural production......................... 283

Table 7.18: Amount, interest, source and
purpose of overdue debts in the 2017/18 season..................... 284

Table 7.19: Area under primary crop production...................... 291

Table 7.20: Production levels: Income from
agricultural production 294

Table 7.21: Percentage ownership of productive
assets in Goromonzi, 2011 & 2018 301

Table 7.22: Percentage ownership of
on-farm infrastructure by peasant cluster (2017/18)................ 302

Table 7.23: Sources of income to fund
asset and infrastructure accumulation, 2016-2018................... 303

Table 7.24: Gender composition, by cooperative
and by peasant classes.. 305

Table 7.25: Access to marketing information by gender........... 306

Table 7.26: Performance of women and men
on the cooperative... 307

Table 7.27: Challenges facing the cooperative

organisations by the peasant class.............................. 308
Table 8.1: Summary list of challenges
observed in Goromonzi cooperatives.......................... 315
Table 8.2: New three-tier membership structure...................... 329

Figures

Figure 2.1: Linkages between the
agrarian question and the cooperative movement.................... 25
Figure 2.2: Development of Hayami's CMS framework.......... 28
Figure 2.3: Evolution of neo-classical cooperative
theory in the 20th century... 42
Figure 2.4: Cook's lifecycle of a cooperative.......................... 44
Figure 2.5: Advantages of adopting producer cooperatives..... 59
Figure 3.1: The rice distribution system during
the era of the Food Control System................................ 75
Figure 3.2: Socio-political periodization
of Japan (1600-2020) 83
Figure 3.3: Pre and post land reform
agrarian structure (1941 – 1955). 89
Figure 3.4: The structure of the new
MAFF internal structure 96
Figure 3.5: Flow of agricultural funds
within the JA cooperative banking system (2019) 98
Figure 3.6: The three tier-structure of
Japanese Agricultural Cooperatives (JA)....................... 102
Figure 3.7: Sources of profits for the
JA-Zenchu (1970-2012).. 107
Figure 4.1: Lending to Agriculture from
Zimbabwe commercial banks (USD)............................. 143
Figure 4.2: Structure of the Zimbabwe
Agricultural Cooperatives Movement 147
Figure 5.1: Matrix code frequency query of themes............... 156
Figure 5.2: Cluster analysis of themes based
on word similarity... 157
Figure 5.3: Number of local agricultural
cooperatives in the JA-GO region. 166
Figure 5.4: Employment trends in

agriculture (%) (1970-2014) 197

Figure 6.1: The most frequently used
words by the farmers 208

Figure 6.2: Number of times that the
issue was mentioned in interviews and FGDs 209

Figure 6.3: Clusters of the relationships
between themes based on word similarities............................... 210

Figure 6.4: Challenges and constraints faced
by farmers during agricultural production.................................. 241

Figure 6.5: Agricultural commodities and the
marketing challenges faced 2015-2018.. 243

Figure 7.3: Two-step cluster model summary 254

Figure 7.4: Cluster composition and characteristic................... 255

Figure 7.5: Custom pivot tables, Chi-square test
and comparison of proportions analysis 258

Figure 7.6: Level of education attained,
and formal agricultural training received 261

Figure 7.7: Reason for forming cooperatives............................. 262

Figure 7.8: Benefits experienced after joining cooperative 265

Figure 7.9: Perception of women integration
and composition in cooperative committees............................... 269

Figure 7.10: Ranking of management by the
cooperative members... 270

Figure 7.11: Access and utilisation of agricultural
production inputs for the 2017/18 season.................................. 286

Figure 7.12: Sources of agricultural inputs
in the 2017/18 agricultural season.. 287

Figure 7.13: Sources of income to purchase
agricultural inputs for the 2017/18 season................................ 288

Figure 7.14: Labour input access and utilisation:
Averages for 2014 to 2018 season... 289

Figure 7.15: Land utilisation rates by significant
crops for the 2017/18 agricultural season 292

Figure 7.16: Top three most utilised marketing
channels by cluster for the 2017/18 season 295

Figure 7.17: Top three most utilised marketing
channels by cooperative (2017/18).. 296

Figure 7.18: Reasons for selecting marketing
channel by cooperative for the 2017/18 season 297
Figure 7.19: Sources of income for cooperative
members 2016-2018 agricultural seasons.................................... 299
Figure 7.20: Average income earnings from
various sources in the 2017 production year.............................. 300
Figure 8.1: Gradual development from a
federated to a mixed cooperative structure 330
Figure 8.2: Development from Single to
Multipurpose cooperatives.. 332
Figure 8.3: New Zimbabwe Cooperative Structure.................... 339

Pictures
Picture 1.1: Field interviews with a small-scale
cooperative member, Chiba, Japan, 2019.................................... 22
Picture 5.1: Cooperative warehouse for
receiving and dispatching farmer's produce 162
Picture 5.2: Produce is associated to a farmer
in the cooperative stand in the supermarket 163
Picture 5.3: Participatory field research, planting
activities and price negotiation meetings 167
Picture 5.4: Round-up herbicide of Monsanto/Bayer
in partnership with JA cooperative.. 168

Textboxes
Textbox 5.1: The story of Nakamura san from
Sanbu Yasai Network.. 201
Textbox 6.1: The story of Tagarika
Irrigation Cooperative 246
Textbox 6.2: The story of Xanadu
A Tractor Cooperative... 247

List of acronyms

A1	Small-Scale land resettlement model
A2	Large-Scale land resettlement model
ACC	Agricultural Cooperative Credit System
AFC	Agricultural Finance Corporation
AFFFC	Agricultural Forestry and Fisheries Finance Corporation
AGM	Annual General Meeting
AQ	Agrarian Question
AREX	Agricultural Research and Extension
CA	Communal Areas
CACU	Central Association of Cooperative Unions (Zimbabwe)
CBOs	Community Based Organisations
CCB	Central Cooperative Bank
CFU	Commercial Farmers Union
CI	Confidence Interval (in statistics)
CSOs	Civil Society Organisations
CS-Pro	Census and Survey Processing System
CUAC	Central Union of Agricultural Cooperative (Japan)
ERS	Export Retention Scheme
ESAP	Economic Structural Adjustment Program
EU	European Union
FAO	Food and Agricultural Organisation
FDGs	Focus Group Discussions
FDI	Foreign Direct Investments
FILP	Fiscal Investment and Loan Program
FOB	Freight on Board
FTLRP	Fast Track Land Reform Program

GDP	Gross Domestic Product
GHQ	General Headquarters
GMB	Grain Marketing Board
GMO	Genetically Modified Organism
GoZ	Government of Zimbabwe
GRDC	Goromonzi Rural District Council
HQ	Headquarters
ICA	International Cooperative Alliance
ICAO	International Cooperative for Agricultural Organisations
ICBA	International Cooperative Banking Association
ICBC	International Cooperative Banking Committee
ICMIF	International Cooperative Mutual Insurance Federation
ILO	International Labour Organisation
IMF	International Monetary Fund
JA	Japanese Agricultural Cooperatives
KMO	Keiser Meyer Olkin (statistical test)
LCF	Life Cycle Framework (for cooperatives)
LHC	Lancaster House Conference
LMF	Labour Managed Firms
LSCF	Large Scale Commercial Farms
MAFF	Ministry of Agriculture, Forestry and Fisheries
MAMID	Ministry of Agriculture Mechanization and Irrigation Development
MNC	Multi-National Company
NFACA	National Federation of Agricultural Cooperative Associations
NFAZ	National Farmer Associations of Zimbabwe
NGOs	Non-Governmental Organisations

NMIFAC	National Mutual Insurance Federation of Agricultural Cooperatives
NR (I-V)	Natural Regions (one to five)
NVIVO	A Qualitative Data Analysis (QDA) computer software package
NWFAC	National Federation of Agricultural Cooperatives
OCCZIM	Organisation of Collective Cooperatives of Zimbabwe
OILS	Open Import License Scheme
OTFM	Organisation of Temporary Farmland Management
PCA	Principal Component Analysis
PPP	Public-Private Partnerships
SACCOs	Savings and Credit Cooperatives
SCAP	Supreme Command of the Allied Powers
SMAIAS	Sam Moyo African Institute for Agrarian Studies
SME	Small-Medium Enterprise
SMECD	Small-Medium Enterprises and Community Development
SMS	Short Message Service
SPSS	Statistical Package for Social Sciences
SSCF	Small-Scale Commercial Farms
TPP	Trans-Pacific Partnership
TV	Television
UDI	Unilateral Declaration of Independence
UK	United Kingdom
USA	United States of America
USD	United States of American Dollars
WB	World Bank
WMF	Worker Managed Firms
WTO	World Trade Organisation

WVA	War Veterans Association
WWI	World War One
WWII	World War Two
ZANU PF	Zimbabwe African National Union-Patriotic Front
ZFU	Zimbabwe Farmers Union
ZIMACE	Zimbabwe Agricultural Commodity Exchange system
ZNFU	Zimbabwe National Farmers Union
ZNWVA	Zimbabwe National War Veterans Association

Map 1: Map of Japan showing the three research sites

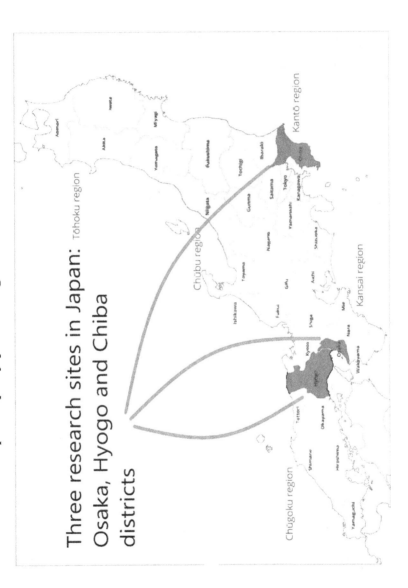

Three research sites in Japan: Osaka, Hyogo and Chiba districts

Map 2: Research sites in Japan

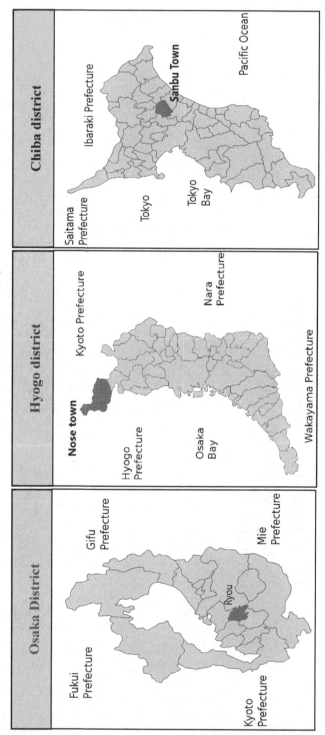

Osaka District

Fukui Prefecture

Gifu Prefecture

Kyoto Prefecture

Ryou

Mie Prefecture

Hyogo district

Nose town

Hyogo Prefecture

Osaka Bay

Kyoto Prefecture

Nara Prefecture

Wakayama Prefecture

Chiba district

Saitama Prefecture

Ibaraki Prefecture

Tokyo

Sanbu Town

Tokyo Bay

Pacific Ocean

xxiv

Map 3: Map of Goromonzi district, Mash-east province, Zimbabwe

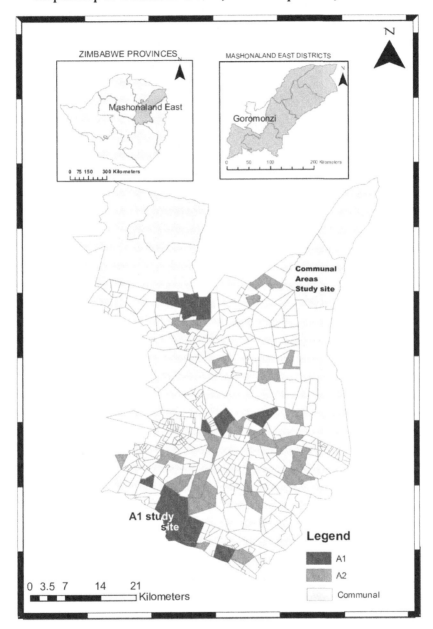

Source: Ministry of Lands, Agriculture and Rural Resettlement (2012)

Chapter One

Introduction to Rural Zimbabwe and its challenges

If you want to go fast, go alone. If you want to go far, go together.
— African Proverb
A single arrow is easily broken, but not three in a bundle.
— Japanese Proverb

My grandfather and the agrarian issues in the 1980s

When I was born in 1987, stagnation of the Zimbabwe economy became more evident despite growth rates of 7-10% recorded since 1980. My father later learnt that the stagnation was caused by a combination of unforeseen socio-political and climatic factors. My parents, like all the 73% of Zimbabweans residing in the rural areas at that time, had lived through the 1982 to 1985 consecutive draughts. They put more blame on the erratic rainfall patterns experienced in Communal Areas (CA) for the devastating misfortune in maize and groundnut yields. Socio-political implications on the sluggish growth rates were imperceptible to them at that time. During this period, approximately 70,000 people of direct European descent held over 70% of fertile agricultural land located in the rainfall-rich portions of the country. And about a hundred times more black-indigenous people controlled 15% of the low-quality land.

My grandfather narrated stories from these times during various visits I made to his homestead in Hurungwe district, Mashonaland West Province. I was told how people resorted to collective action in times of stress as they organised collaborative input acquisition, group cultivation, ploughing (land preparation) partnerships, weeding and harvesting parties, communal transportation facilities and community gardens (mainly operated by women). My grandfather, although one of the most hardworking men I know, was a superstitious man. In late September 1985, he fled from Zvimba CA and moved to Hurungwe CA after suspecting that witches were

1

after him and his family. In one of many fireside stories, he narrated to me how he had arrived in Hurungwe, although initially treated as a *Zezuru* foreigner in the *Gova/Korekore* tribal lands, he was momentarily welcomed by his neighbours, the village-head (*Sabhuku*) and the spiritual leaders (*maSvikiro*). They eventually showed him Hurungwe rural life.

One of the community leaders leased to him (for free), a piece of land to plough because his land was not yet cleared. He attributed his eventual success in agricultural production to the informal labour and production groups in Hurungwe which also consisted of various farmer collective action activities in input/output marketing and irrigation systems of horticultural crops (*paprika*) and cash crops (cotton). Reflecting on these stories in my adult life reminds me of how collective action has always saved communal area Zimbabweans from harsh economic shocks.

To curtail the stagnation of the mid-1980s, the Zimbabwe government implemented import substitution policies, albeit with little success. The 1987-90 economic stagnation eventually drove them to adopt the Economic Structural Adjustment Programmes (ESAP) as advised by the World Bank (WB) and International Monetary Fund (IMF) in 1990. Unknown to my grandfather, this policy would have devastating effects on his life and that of his rural relatives and friends. While some welcomed the program highlighting that it was being implemented in other countries such as Kenya, Zambia and Tanzania, my grandfather did not like it. His dislike for the program was neither ideological nor based on some technical experience in policy formulation. But grandfather did not understand why the government was abruptly reducing grain silo construction projects in the countryside, or why the District Development Fund (DDF) was scaling down on construction and maintenance of rural roads. He was shocked to understand that the Cold Storage Commission (CSC) was scaling down on its frequent visits to the local growth point to purchase cattle from the farmers. He did not like ESAP because it meant that he had to look for input and output markets on his own, which was previously easier to get with the help of the state.

Due to the poor prospects for gainful employment in CA agriculture (e.g., Zvimba and Hurungwe), many young people of my father's age moved to the capital city, Harare, to seek for greener pastures at the beginning of the 1990s. However, by 1998, it was clear that the promise made through ESAP was but a fallacy. By this time, leaders of rural movements, with the help of scholars had identified the stumbling block to rural transformation as the unequal land ownership patterns which left the majority owning marginal lands, and unable to reproduce themselves. The demand for land reform was intensifying. My grandfather felt that he was too old to apply for land, the prospects of relocating again deterred his participation. My father, who had moved to the city, thought that the reform program was for the peasants and that it offered little for the proletariat. It was my uncle (older brother to my father), who joined the Zimbabwe War Veterans Associations (ZWV) to push the government for land redistribution. At that time, uncle's only access to land was through sub-divisions from grandfather and occasional land renting, a situation peculiar to many rural young men. As the reform progressed, even my father, just like many others in the urban areas, realised the importance of land reform and joined the movement to seek for a piece of land.

Around 2000, the pressure from people such as my uncle in rural areas, and my father in the city intensified. President Robert Mugabe eventually signed into law, the Fast-Track Land Reform Programme (FTLRP) which sought to correct the unbalanced land ownership patterns in the rural areas through redistribution. Unknown to my teenage self (I was 13 years old then), this policy would affect a significant part of my adult life. The FTLRP ignited my love for agrarian issues so much that I became an agrarian scholar. The reform increased the number of small-scale farming households by 300,000 while expanding their total landholding by approximately 10% (Moyo, 2011c, p. 512, 2015, p. 43). Because my uncle was on the front line, he received an A1[1]-self-contained farm, and my father (who only participated from the city) did not. I used to visit my

---

[1] An A1-self-contained farm is a small-scale farmer who received land during the 2000 FTLRP, the farmer owns individual arable, grazing and homestead area ranging from 10 to 50 hectares depending on agro-ecological region.

uncle's new farm during my late high school and undergraduate university holidays. While the FTLRP had improved access to land and natural resources, there remained fundamental issues of access to input and output markets that my uncle (a tobacco farmer) faced every season.

Of course, I could not understand several of the agrarian issues while in high school. However, it became apparent during my university days, that the challenges faced by farmers emanated mainly from the lack of socio-economic structural organisation when the farmers engaged with state policy or market mechanisms. The private sector had retreated after the FTLRP, and the government was financially outstretched. The increased number of peasants amid an under-resourced state and a retreating private sector necessitated the formation of robust social organisations to enable the farmers to deal with various production challenges. Improved social institutions were necessary to give farmers a voice, to strengthen their agency within such interactions as Community-Market-State (CMS – page 26).

In this book, through the support of literature, theory, and empirical data analysis, I submit that small-scale production remains one of the critical sectors that ensure food sovereignty and food security. Small-scale farmers help resolve a substantial portion of the agrarian question (AQ) and hence land reforms (radical or otherwise) should aggressively include the small-scale farmers, not as subjects of rural development, but as active stakeholders of the processes. I further argue that for a long time, small-scale participation has been hindered by low levels structural organisation of the farmers at local level, which has left them vulnerable to exploitation from the markets, the state, or very often, a combination of both. Structural organisation can be achieved through collective action mechanism such as agricultural cooperatives. Thus, the central objective of this book is to examine the current and potential role of agricultural cooperatives[2] in agrarian development. I wanted to understand how

---

[2] A cooperative is a group of autonomous persons, voluntarily joined in a democratically controlled association to meet a common economic, political, social, and cultural goals (agricultural production for the case of agricultural cooperative).

cooperatives can transform the socio-economic lives of my grandfather, my uncle, my father (and his children) who represent the communal area, the resettled farmer, the urban workers (and the future) respectively.

The book, therefore, seeks to achieve this by examining the challenges faced in the Japanese agricultural cooperative system, and how collective action helped Japan to overcome her challenges. Japan had a radical land reform closely resembling that of Zimbabwe and understanding how she delt with the resulting agrarian questions was attractive to me and inspired the title of this book. Using the lessons from Japan, the book then produced a new cooperative model for Zimbabwe. The central message is, therefore, that for developing countries like Zimbabwe which have completed their land reforms, i) small-scale production is vital, ii) these small-scale producers should be united through robust socio-economic organisations, and iii) the socio-economic organisations should take the form of agricultural cooperatives.

The classic and contemporary AQ, from Japan to Zimbabwe

When I was younger, I relied entirely on my grandfather's wisdom to explain rural socio-economic issues in Zimbabwe. Even though I became aware of theoretical approaches to understanding these issues after attaining my first degree from the University of Zimbabwe, his description of rural politics and social challenges, although riddled with superstition and religious undertones, reflected many aspects of issues discussed in the classic and contemporary agrarian question theories. The agrarian question is defined as a series of questions on how agrarian societies (rural areas) can be developed and moved to industrial or modern societies (post-agrarian) (Moyo & Yeros, 2005b, p. 14). It asks what would be the best way that the government or the community can restructure agricultural labour, financing, and commodity markets as a way to move away from backwardness. My grandfather knew that Africa's developing economies are heavily reliant on agricultural production for

---

Its mission is to protect the interest of its members (Chayanov, 1991; Cotterill, 1988; Hoyt, 1996; ILO, 2002; Mhembwe & Dube, 2017; Yamashita, 2015a)

employment, generation of foreign exchange and food security because of the dominant rural populations (60-70%) (World Bank, 2017, 2019b). Farms were sites of both production and consumption. Although large-scale farmers existed, their numbers were limited in African agrarian structures due to a blocked agrarian transition (see Amin, 2012, pp. 18–21). On the other hand, family farms or peasants were found in abundance and faced socio-economic challenges such as food insecurity and malnutrition. These peasant farmers are unable to access input markets, which directly contribute to low productivity and low outputs for the agricultural markets.

The problem of rural poverty and underdevelopment threatens not just my marginalised grandfather, but the nation itself and human civilisation in general (Moyo et al., 2013, p. 94). Before the 1980s, peasants dominated most societies because significant portions of the population relied on agriculture as a livelihood. Those societies that 'moved' from peasant to post-peasant (mostly in the Global North) did so by resolving various forms of their classical AQ, either through the industrialisation path or the peasant path. Current efforts to move peasant societies in the Global South have often involved choosing between the said two AQ-resolving approaches. What then is the best way of developing the rural areas and solving the AQ in the Global South, in Africa and Zimbabwe? How can we advance a development trajectory for my rural relatives that protects them from the vagaries of capitalist exploitation[3] and ensures that they gain control of their rural economies? This is the agrarian question which even my grandfather often unknowingly discussed with my teenage self.

Development in the North (America, Europe, and Japan) through industrialisation was historically subsidised by the peasantry in those countries as well as in the developing countries in the South via colonies. The exploitation through global value chains continues till this very day (Amin, 2012, pp. 12–13; Chayanov, 1991; Moyo et al., 2013, pp. 103–105; Patnaik et al., 2011). Some empirical evidence points that, over the last century, flirtations with neo-liberal

---

[3] Capitalist exploitation is unpaid surplus labour and occurs when the price of labour on the market is significantly lower than its actual value. The surplus value (value of labour minus price of labour) will accrue to (or be exploited by) the capitalist (Lenin, 1921; Marx, 1992, pp. 320–330).

frameworks have resulted in the propagation of social and economic differentiation at both national and global levels.

Some elements of the contemporary AQ

The classical AQ has often been conceptualised as a struggle to move farmers out of economic backwardness through whatever means necessary (Byres, 1995); e.g. through a resolution the AQ of capital and labour (Bernstein, 2004, pp. 200–205). The contemporary AQ on the other hand, as Moyo et al. (2013, pp. 94–103) put it, 'restores national sovereignty to its proper place in the classical AQ' (AQ of liberation). The restoration of the national question into the classic AQ opens up spaces for other sub-questions (such as the gender and ecology questions). Efforts to increase agricultural production have to be cognizant of international trading equilibriums, global production, market forces, gender dynamics, environment, and the ubiquitous hegemony of transnational capital. Peasants cannot avoid markets altogether; thus, they need to find ways to safeguard themselves upon interaction.

As my grandfather could rightly attest to, in Zimbabwe (or Africa), post-independence policies such as import substitution, ESAP or market liberalisation resulted in the extension of the hegemony of capital in the rural economy (Shivji, 2009). Recent approaches facilitated by market liberalisation include out-grower schemes, contract farming, insurance, patented technologies (chemicals and suicide seeds) and provision of credit at usury, these have picked up from where ESAP left (Mazwi & Muchetu, 2015, p. 18; McMichael, 2013, pp. 671–673). Global events such as the food crisis of 2007/8 led to increased 'land grabs' and widespread pauperisation of the majority in the Global South, hence highlight the problem at the global level (Amin, 2012, pp. 18–20; McMichael, 2013, p. 671, 2014, p. 10). Increasingly, peasant families across Asia, Africa and South America are unable to reproduce themselves.

All these socio-economic ills are the basis of the contemporary AQ in Africa, which led some scholars (Amin, 2012; La Via Campesina, 2000; McMichael, 2007, p. 415) to call for a total global disengagement from the markets. They argue that capitalism cannot be expected to solve global inequality problems. After all, in nature, it is a system that is socially inefficient because it simultaneously propagates over-production & under-production, over-consumption & under-consumption, the 'stuffed' & the starved, and hence can never resolve its contradictions (Akram-Lodhi & Kay, 2010, pp. 177–179; Amin, 2012, pp. 12–15; Shanin, 1981). Thus, the pervasiveness of contemporary financial capitalism and its continued penetration of the peasant economies transcends to a negative impact on rural lives. That is the complex nature of the classical AQ.

However, in trying to solve the AQ, it is noteworthy that a full disengagement from the capitalist mode of production is a herculean task and may not be desirable. What is required is the establishment of genuine, fair, and equal platforms in which all the three actors of the economy (the state, community, and the private sector) can extract fair value from their interaction in the market. Given the hegemony of both the government and the private sector, the poor majority population needs to unite against all institutions that perpetuate exploitation and to free themselves through decommodification of land and labour. Thus, as I often discussed with my uncle at his resettled farm in Nyamapipi, Mashonaland West province, the state should resolve the unprecedented increase in the volatility of input/output markets, unfair trade practices (domination of agricultural markets by MNCs), climate change, environmental degradation and gender inequalities that restrict the development of the rural areas. What needs to be done is simple; the method to achieve this is the biggest issue.

Agrarian questions in the north: Contemporary agrarian issues in Japan

The agrarian question is not static *per se* but evolves; it may differ between the Global North and the Global South in scope and sequences. Even in areas such as rural Japan, the peasantry still exists, holding tiny pieces of defragmented land, and utilising mainly own family labour. These types of peasants differ from those of the Global South in their utilisation of latest technology and machinery (I called them *mechanised peasants*). Moyo, Jha, & Yeros (2013, p. 94) argued that the AQ in the Global North remains unresolved, and most literature and scholarly work that concludes otherwise adopted

8

an overly reductionist framework to the meaning and definitions of the AQ.

In Japan, food sovereignty issues have recently sparked agricultural debates as the country now depends on rice imports which is often justified by the fact that agricultural production accounts for less than 2% of the GDP, and hence should not be protected (Ashkenazi & Jacob, 2003, p. 17; Esham et al., 2012; Yamashita, 2015b). On the contrary, protectionist scholars argue that the countryside is a site of rural reproduction, cultural-symbolism and national environmental reproduction (Ishida, 2003). This intellection has profound and overarching implications even for development discourses, and policy recommendations developed from the Global North and intended for the Global South. Consequently, ideas of following the European path to development is problematic for the case of Africa. This is because Africa does not have the options to grab territories and extract labour from other countries, as was the case for the Global North (Moyo et al., 2013, p. 94). Africa needs an alternative path *of*, *by* and *for* the peasantry as provided by cooperatives, I argue.

It is imperative to understand the agrarian question both in Japan (in the context of the post-war (1945-1948) land reform) and in Zimbabwe in the context of the recent FTLRP which exponentially re-peasantized the Zimbabwean agrarian structure (Moyo, 2011c, p. 513; Moyo & Yeros, 2013, p. 342; Scoones et al., 2010, p. 232).

So, what is the solution: Through large farms or the peasant path?

The term 'Agrarian Question' was too abstract, not just for my grandfather, but the majority of his other rural counterparts (and sometimes they are too abstract for scholars too). However, AQ asks whether rural areas should be developed by aggregating land and resources (on a national/global scale) to a few large-scale 'efficient' food producers or should small-scale production be pursued instead? Those that favour large-scale development highlight that the average ratio of productivity in the Global North to that of the Global South is approaching 500:1 (Mazoyer & Roudart, 2006, p. 451). The proposition by such organisations as the World Trade Organisation

9

(WTO), based on comparative advantage and economic liberalisation theories, is that those with access to financial capital should take over agricultural production from those without (Collier & Dercon, 2014; McMichael, 2007, pp. 408–409). Therefore, world food security can be achieved by aggregating land under all small-scale producers to approximately 20 million larger-scale producers. This seems to make sense, but what would then happen to the three billion peasants currently engaged in auto-consumption production on the said land? Will they become casual labour in the new large-scale farms as asked by McMichael (2007, p. 409), or will this path lead to a genocide of almost half of humanity since not more than 7% of the peasants can be absorbed into labour as upheld by Amin (2012, pp. 13–14).

My position in this book is that scholarship and multi-lateral organisations must move away from any national development trajectories that exclude small-scale production. Jayne *et al.* (2019, pp. 1–2) acknowledge the possibility of agricultural growth without family farms as in the Latin American *Latifundio*. However, the transition does not move millions of family farmers out of poverty. As evidenced through the multi-lateral drive to support family farms in the past couple of decades, think-tanks and multi-lateral development agencies (WB and Food and Agriculture Organisation – FAO) now realise that small-scale agriculture is a crucial component for the African development process (FAO & IFAD, 2019).

African governments had started to seriously consider the small-scale route, evidenced by policies that aim to redistribute production resources, especially land (Moyo et al., 2013, p. 97). As the realities of inescapable land redistribution settle across most Southern Africa, the debate has moved from 'whether it would be necessary to carry out redistributive land reforms', to a more nuanced conversation of 'what would be the best way of carrying out the redistribution of the land' as exemplified by South Africa. Small-scale farmers are ideal to kick-start a cycle of local spending as compared to urbanites. Local spending encourages movement of money within the local economy which significantly boosts demand for goods locally leading to a virtuous cycle in which the rural and urban labour force is a market for each other (Jayne et al., 2019, pp. 1–3; Johnston & Kilby, 1975).

So how can governments and policymakers achieve this? Several scholarly solutions have focused on trying to correct or remedy market access, for example, contract farming, government input subsidies and mechanisation schemes (FAO & AUC, 2018; Little & Watts, 1994; Moyo et al., 2014). Others, on the other hand, advocate for gross policy reorientation through structural transformation[4]. Badiane (1997), Leavy & Poulton (2008, p. 5) and Wiggins et al. (2011, pp. 57–59) have pointed out that under ideal conditions mainly put in place by the government; the small-scale farmers can actively take part and extract benefit from global markets. While such channels as contract farming, out-grower farming, commercialisation and provision of subsidies were tried on various cash & non-cash crops, they still expose the farmer to the vagaries of capitalist exploitation (Little & Watts, 1994; Mazwi & Muchetu, 2015). This is because financial institutions, the input and output markets are owned or controlled by capital. However, if these institutions and these markets are placed under the control of the farmers themselves, then the farmers can reap a fair share from the agricultural value chains. Cooperativism, therefore, has the potential to transform power and control relations in the rural economy as they did in Asian (Japanese) agriculture.

'Is it possible to compare Zimbabwe and Japan?'

The reader of this book might be wondering how and why someone would try to relate the agrarian history of Japan and to that of Zimbabwe. Or whether it is possible and even desirable to do so. First let me clarify that, while I might have utilised a comparative approach, this book is for Zimbabwe. Secondly, while the two countries are in different spatial and temporal locations with different cultures, as shall become evident as the reader proceeds, there are several insights one can get from Japan's experience and not

---

[4] Structural transformation is a development process achieved by restructuring production resources from primary industries to secondary industries. It occurs mainly when agriculture's contributions to the economy grows smaller despite increases in its overall production and is pronounced by stimulated processes of urbanisation (Timmer, 2009, p. 5).

anywhere else. There are several uncannily similar resemblances in policy formulation and socio-economic characteristics between Zimbabwe (Africa) and Japan (Asia). The radical nature of both country land reforms, and the respective grain policies became the basis for comparison.

I decided to study Japan and draw lessons from her experiences with the AQ mainly because Japan carried out radical but complete land reform. It also adopted a series of robust post-reform rural community development policies that propelled the agricultural sector during the 'golden era'. For developing countries like Zimbabwe, whose agrarian trajectory seems to be where countries in Asia were just before the Asian Green Revolution, it has become paramount for scholars to draw lessons from their experiences. This is because the nature of the AQ in post-reform Zimbabwe has notably hinged on the inability to remove information asymmetries and market imperfections [5] as well as market misconduct which results in exploitation, underdevelopment, and poverty. In Japan, capitalism had fully established itself by the time of the land reform in 1945. Peasant cooperatives provided a base and a quasi-arm for government to engage with the community and the private sector to develop the countryside. Thus, learning from Japan's experience after its land reforms becomes a virtue.

After arriving in Japan to pursue my PhD in April 2017, I volunteered to work on a farm in Nose village, Hyogo district, where I learned a lot about the trajectory that Japanese agriculture had taken since their land reform. I remember every night we would watch a documentary about cooperatives or agricultural philosophies, or I would listen to Tohira Kazuo describe how the face of agriculture had stood firm despite various attacks to liberalise up to the beginning of the 1990s. Tohira spoke perfect English and had so much knowledge about the cooperative movement, the national and international political economy. It was at this farm that I realised how cooperatives were at the realm of everything agriculture from 1945

---

[5] Information asymmetries occur during transactions and where one party has more information (sometimes wrong information) than the other (see Chapter Two, page 26) (Stiglitz, 2002, p. 470).

until the late 2000s. I was fascinated and thus motivated to study the Japanese Agricultural Cooperative systems (JA).

A scan through the Japanese agricultural literature reveals an overwhelming pervasiveness of the Japanese Agricultural Cooperatives (Nōkyō - 農協 or JA). The cooperative establishment was an active policy adopted after WWII to protect rural farmers against the vagaries of capitalism. These helped to launch structured social organisations that would make the implementation of the agricultural policy easier for the government as well as for the safe farmer-market inter-connection. The JA represented the farmers in the negotiations between the farmers, the government, and the market (local, national, and international). The negotiations mainly kept rice prices up to secure its continued production by the farmers, as well as the protection and subsidisation of the JA activities. The JA formed the basis for rural-based development that saw it rise to become one of the most successful cooperatives in the world. Today the JA's annual income is in the top three income earners, not just among cooperatives or government institutions, but including private companies in the country and the world's top 40 (Esham et al., 2012, p. 943; Godo, 2014).

On the other hand, Southern African land reforms have either been tough to implement (titling in Kenya, restitution in South Africa) or have faced endogenous and exogenous post-reform challenges (Zimbabwe). Therefore, the herculean task for scholars has been to try to understand how countries in Asia managed to record such levels of success with their land reforms, and most importantly how they managed to comprehend and deal with the resultant agrarian structure. Despite the differences between Japanese and Zimbabwean (or any African country) land reforms, it remains critical to draw lessons on how Japan dealt with the realities resulting from the creation of numerous small-holder farmers in a short space of time. The safe integration and active participation of smallholder farmers into the volatile globalised markets (opposing adverse incorporation) while producing adequate food for the home market is of concern for the governments (Hickey & du Toit, 2007, p. 5; La Via Campesina, 2000). Considering the increased number of family farms, and also the accelerated rate of technological and financial diffusion to the peasantry, protecting them from the jaws of capitalism has become indispensable. As already mentioned, this process needs to be critically managed because failure will result in negative consequences on small-scale farmers, as shall be argued in Chapter Two.

The Siamese twins: The State and the Market in land markets

By 2000, the relationship between the GoZ and the market (especially land markets) had become cosy given the level of mutual suspicion in use of dishonesty to benefit from land market transaction (detailed discussion in Chapter Four). The GoZ after succumbing to pressure from below embarked on a radical land reform programme that saw the redistribution of approximately 10 million hectares of prime land to over 169,000 households between 2000 and 2010 (Moyo, 2011c, p. 496). This effectively opened up land access to the farmers, thus providing one of the fundamental requirements for agricultural development (Moyo, 2005b, p. 146, 2011b, p. 261). However, the state has struggled to support the new agrarian structure (input and output markets). Throughout Sam Moyo's work on land reforms, he declared that market-led land reforms, though they respect private property rights, were slow and would never result in socially or racially equitable land ownership patterns (Moyo, 1992, 2007, 2015). On the other hand, land reforms that take a more radical nature, despite flouting liberal economic and political rights, results in more equitable land redistribution system (Moyo, 1992, 2007; Muchetu, 2018, pp. 90–91). Furthermore, Moyo noted that land reform was indeed a necessary but not sufficient condition for the emancipation of rural areas from poverty and underdevelopment (Moyo & Matondi, 2004).

The broader macro-economic challenges that engulfed the country after the land reform affected the inputs markets negatively (for seed, fertilisers, herbicides, pesticides, tools and implements) due to the absence of agrarian finance (Mujeyi, 2010, pp. 6–10). These challenges compounded other problems synonymous with post-land reform farm establishment phases such as underdeveloped infrastructure (roads, clinics, schools, farm infrastructure and electricity), underdeveloped social networks, as well as underdeveloped local (output) markets. The challenges are part of the fundamental tenets of contemporary AQ (Moyo et al., 2013).

If the radical bottom-up redistributive land reforms translate to a focus on the peasant path, a way to consolidate the peasants into formal groupings needs to be envisaged. In this respect, as indicated earlier, this book proposes support for the establishment of agricultural cooperatives in the rural areas to enable the consolidation of capital, social networks and development of input and output markets. As shall be discussed in greater details in Chapter Two, several forms of cooperatives can be established in the rural areas, all with different levels of management styles and economic sectors where they can thrive. This book rekindles cooperativism approaches for policymakers and development agents as a conduit for rural development, fighting poverty and inequalities within the peasantry. Because the target is the post-land reform agricultural sector, bloated by an expanded peasantry, I propose the Chayanovian peasant cooperative model.

Cheers to agrarian populism: Chayanovian as an alternative

Some may argue that cooperatives are an old idea not worthy of considering in the 21st century (Akwabi-Ameyaw, 1997, pp. 444, 453). However, countries in the Global North (including Japan) utilised the model during their developmental stages. They were able to reduce rural poverty as seen in the case of first cooperatives in the English Rochdale Pioneers cooperatives and Germany's Raiffeisen & Schulze-Delitzsch cooperatives (Holyoake, 2016; Prakash, 2003). The movement then spread to other areas, albeit with some localised modifications such as the collectives and communes in Russia and China respectively, or the Japan Agricultural Cooperatives (JA) in Japan in the mid-1900s. In recent times, agricultural cooperatives have been successful in China. The trajectory of the Chinese peasantry is not too dissimilar to that of the African peasantry (Hairong & Yiyuan, 2013, p. 964).

Considering that the world economic structure has inclined towards capitalist production over the past 100 years, and also given that, historically, cooperatives were associated with socialism and communism, promotion of cooperatives is feared as re-entry for socialism (see discussions in Dore, 2012; Jossa, 2014; Lenin, 1923; Marx, 1996). Indeed, the growth of most post-independence government-led cooperatives in Africa increased with the rise of the Soviet Union and later with communist China. For Marxism and Leninism, cooperatives were a means to an end and not an end in itself; thus, the cooperative could be a powerful means of reorganizing societies which may lead to socialism.

In their studies of Kenyan and Tanzanian cooperatives, Dondo (2012) utilised a classical economics model to analyse the potential of cooperatives. They concluded that cooperatives were a viable[6] channel for rural development. Ortmann & King (2007) also used the New Institutional Economics (NIE) theory in their study of South African cooperatives. They concluded that cooperatives played a significant role because they supplied the requisites for farming and also were at the fore of commodity marketing. This points out that cooperatives are not against everything that capitalism stands for but are there to remedy the contradictions that it brings. There is need to realise that cooperatives are an important alternative in developing the countryside even within a new institutional economics framework as also believed in the Chinese cooperative movement:

> Cooperation is opposed to capitalism. Its impact will more than undermine economic imperialism. But its approach is different from that of Marxism. Cooperativism does not emphasise revolution but emphasises construction. It does not rely on the state but is based on organisations. Its approach is gradual, and its action is far-reaching. – Chen (1983, 97) in Hairong & Yiyuan (2013, p. 957)

Here lies an acknowledgement that either liberal or socialist fundamentalism should not despise cooperatives; instead, it is a platform for marrying the two. This amalgam between socialist-oriented theorisation and new institutional economics brings us to

---

[6] Viable cooperatives that can be managed efficiently to achieve real grassroots development in the countryside. Viable cooperative should be feasible and capable of working successfully within the contemporary neo-liberal production system.

the works of Alexander Vasilevich Chayanov, a Russian economist of the early 20th century. Realising the omnipresence of capitalism and the fact that peasants cannot escape the markets, he developed a theory of farmer's organisations that is particularly appealing. The NIE theory forces cooperatives to behave like capitalist corporations, while Marxist theory, at worst wanted to use the peasant cooperative to attain state socialism (or at best to ignore them entirely because they were seen as capitalistic in nature). However, Chayanov's cooperative theory takes the form of bottom-up socialism based on individual peasant farmers engaged in auto-consumption while at the same time participating in the agricultural markets. His emphasis on the three facts that i) peasants were not embryonic capitalist, ii) were motivated by use-value, and iii) the household was a site of production and consumption differentiated his theory from others. Agricultural cooperatives by small-scale farmers offer a third-way type of solution to the peasant problems (Chayanov, 1925; Thorner, 1965, p. 229) (discussed in detail in Chapter Two).

Issues and themes

This book is an attempt to conceptualize several struggles affecting my grandfather, my uncle, and several other rural farmers across Zimbabwe whom I have interacted with growing up, during my masters' fieldwork, and the time I worked at the Sam Moyo African Institute for Agrarian Studies (SMAIAS). It is this experience that has shaped the way I conceptualize the problem of the Zimbabwean agrarian spaces. For me, the biggest issue manifest as an unprecedented increase of volatility of input & output markets, unfair trade practices, climate change, environmental degradation and gender inequalities that restrict the development of the rural areas. In the resettled areas, the problems are exacerbated by the non-existent or inadequate infrastructure, which affects access to inputs and outputs markets. Those farmers that have access do so at a considerable cost which is not reflected in the price for their produce. They have limited spaces to negotiate. The farmers are left at the mercy of the markets or as Pinto (2009, p. 3) put it, 'throw the rural organisations into the water and let us see which ones can swim'.

Therefore, the objective of this book is to examine how cooperatives deal with the multi-faceted challenges in Japan & Zimbabwe's post-land reform marketing system. The process involved drawing lessons from experiences of Japan's post-land reform agriculture system in which the JA played a key role. I examined the Japanese post-war agricultural growth, and how the JA navigated the political economy of agricultural production in an increasingly neo-liberal national economic production environment. Given the accelerated efforts to adopt liberal agricultural policies currently underway in Zimbabwe, the experience of Japan became extremely relevant to envisaging a development path for agricultural cooperatives in Zimbabwe.

The main goal was not to compare the experiences of the two countries in isolation (done in the first half of the book), but to go a step further and provide strong practical recommendations for Zimbabwean cooperatives. The second half of the book becomes denser on the description of Zimbabwean agriculture to lay context to practical recommendations for Zimbabwe agriculture. This exercise sought to answer questions such as 'Can cooperativism survive and prosper within a neo-liberal and industrialised national mode of production?' 'What is the effect of post-land reform on the growth and development of cooperatives?' 'What is the state of cooperatives in Zimbabwe, and what are the future opportunities for growth?' 'What role should the state play in the development of cooperatives?' and 'What form of the cooperative framework is ideal for the post-reform cooperative movement in Zimbabwe?'

Approaches and Analysis strategy

Unlike my grandfather, who could describe the challenges and solutions needed for the rural areas based on hands-on experience in the rural areas, I neither had this experience nor could I ever accumulate enough in my lifespan. Thus, I had to rely on grandfathers, uncles, and many other farmers' narratives to discuss the AQ in the countryside and its solutions. The mixed-method data collection process both in Japan and Zimbabwe utilised case study approaches where interviews, Focus Group Discussions (FGDs) and questionnaires were used.

After coming to Japan to pursue my PhD studies, my academic supervisor encouraged me to spend some time in rural Japan to get an understanding of Japanese agrarian issues. Given the nature of rural Japan, characterised by closed societies with a minimum number of people who can speak English, a fair amount of rapport had to be harnessed before being accepted to do fieldwork in these areas. Several phone calls, emails, and lunch meetings later, we were granted permission to stay at a cattle-rearing farm in Nose district. I was subsequently granted permission to spend some time on farms in Chiba district as well as in Ryuo, Shiga district. In these three villages, I learnt how cooperatives were at the centre of rural production and reproduction, and this motivated me to adopt a kind of 'soft-comparative' analytical approach. I called it *soft* because I limited the analysis to 'learning the experiences of the Japanese cooperatives' to develop a new cooperative model for Zimbabwe as the final output. A comparative study (in the strict sense of the term) would have required more financial and time resources which were not at my disposal.

Through the multi-case study approach, my arguments relied on data collected in six villages. There were three villages in Zimbabwe – Tagarika, Xanadu and Juru (all found in Goromonzi districts – see Map 3); and three in Japan – Nose (Hyogo district), Ryuo (Osaka districts), and Sanbu (Chiba district) – see Map 1 & 2. I decided to study these areas in the hope of finding different types of cooperative associations. For example, Tagarika and Xanadu A are villages located in the resettled areas while Juru village is found in the communal areas established during the colonial era. Goromonzi is one of the 59 districts found in one of the ten Zimbabwe provinces. It is predominantly in natural regions II (NRII [7]). Although the district is close to Harare, and hence is classified as peri-urban, 75%

---

[7] Zimbabwe is divided into five agro-ecological/natural regions (NR) predominantly depending on rainfall, temperatures and land-use patterns. They range from NRI – the wettest region, has moderate temperatures and is conducive for intensive crop production; to NRV – the driest, hottest and is conducive for livestock rearing (Vincent & Thomas, 1960).

of its population is rural (44.3% of these live in the commercial farming areas- LSCF, SSCF, A1 and A2). CA settlers rely on rain-fed agriculture, while commercial spaces have a mixture of rainfed and irrigation facilities. The A1 and A2 are the new farmers who got land in the 2000-2003 FTLRP. The experience I gained while working at the SMAIS was instrumental in shaping the methodology for collecting Zimbabwe data.

The pattern of land use and ownership in Goromonzi has undergone several changes in the past two decades. As it stands, a combination of freehold and state-owned characterises its land tenure. LSCF, SSCF areas and resettled areas all fall under the state-owned land while urbanised areas in the district are under freehold tenure. The FTLRP resulted in three different types of agrarian agricultural producers, small-scale CA, small-to-medium scale A1, and large-scale sized A2 farms. In terms of numbers nationwide, most of the households obtained small-to-medium scale farms, and it is these that my study focused. There are about 12 A1 farms and 2 CA sites in Goromonzi.

Approximately 50% of research participants were from CA, where farmers hold 0.1-1ha of arable land and the rest were from newly resettled areas (A1), where farmers have an average of 6ha of arable land. *Table 1.1* summarizes geographical and essential socio-economic characteristics of the study areas. The CA is formerly known as the Native Reserve areas in which black people settled during colonial times, and it is the place where the Rhodesian government had established the old cooperatives.

During various trips to Zimbabwe between 2017 and 2020, I interviewed the current, and former cooperative registrar, assistant registrars, co-operators, cooperative leaders, cooperative experts, chairperson of CACU[8], and some ministry officials.

---

[8] The Central Association of Cooperative Union (CACU) is the Apex body responsible for representing all cooperatives at the national and regional levels. Its headquarters are in Msasa, Harare and is currently chaired by Mrs Maziva. It is the Zimbabwe's equivalent of the JA-Zennoh in Japan.

Table 1.1: Some background information for the research sites

	Zimbabwe			Japan		
Research site	Tagarika	Xanadu	Juru	Nose	Ryuo	Sanbu
Prefecture	Goromonzi			Hyogo	Shiga	Chiba
Area (ha)	25,407,200			839,600	401,700	27,177
Total pop	224,987			10,072	11,996	977,247
% Females	50.5			57	48.1	
No. HHHs	56,246			4,539	4417	426,765
Avg per HH	4			2.2	2.7	2.3
Rainfall	900			1,444	1,657	1,637
Economy	Agriculture			Agriculture	Industry	Industry & service
Temperature	17.5			15.9	14.5	16.3
Topography	Flat fields			Mountainous	Flat fields	Flat fields
Avg land size	6		1	< 0.5		
Name of coop	Tagarika	Xanadu A	Juru	Yotsuba	JA-Green Omi	Sanbu Yasai Network
Main crop	Maize, Livestock, Horticulture			Chestnuts, rice, charcoal, vegies	rice, vegies	Vegies
Coop focus	-	-	-	Livestock, vegetables	-	-

Source: (Akanemoto, 2016; WikiMedia-Commons, 2006; ZimStats, 2012)

In Japan, on the other hand, Ryuo village is under the typical JA cooperative structure formed after the Japanese land reform of 1945, Sanbu cooperative is a hybrid of the first and those that seek to integrate cooperatives more into the markets, and cooperative structure in Nose was the stuck opposite of the first two. An understanding of these different types of cooperatives would be necessary to compare (or learn from) for the cooperative movements in Zimbabwe.

Choosing a data collection technique and tools for Japan was quite complicated, worse still trying to adopt uniform approaches to enable comparative analysis. Given the limited Japanese language skills I possessed, it made more sense to utilise a structured questionnaire to collect data. However, most of the questions that are easily obtainable through a questionnaire are readily available (though in the Japanese language) on several other online and physical platforms. Thus, to get more in-depth information about the current and trajectory of the Japanese movement, I relied on qualitative techniques (observations, interviews and FGDs) of collecting data.

Picture 1.1: Field interviews with small-scale cooperative members, Chiba, Japan, 2019. Source: Field research photos, the author 2019

Many of the interviews carried out with the actual farmers or members of three different cooperatives had to use structured interview guides to direct the flow of the interviews (*Picture 1.1*). The technique was more flexible than the questionnaire, which needs clarity if respondents are unclear about specific terms. With the help of a translator in some cases, I interviewed knowledgeable stakeholders in the field and history of Japanese land reform and the agricultural cooperative movement. The data analysis process relied on Statistical Package for Social Sciences (SPSS-25) for quantitative data, and QSR International® analysis software and Nvivo® for qualitative data.

Chapter Two

Some Notes on the Theory of Peasant Cooperative

The more impressive a theory is, the greater is the simplicity of its premises, the more different are the kinds of things it relates and the more extended the range of its applicability [...] It is a theory that decides what is observable.

– Albert Einstein

Introduction

It is important to first discuss, from a theoretical perspective, how cooperatives are better placed to resolve the AQ than other social organisations? This book agrees with the arguments advanced by Scott (1998) and Hayami & Godo (2005) who recognized an urgent need to strengthen and amplify voices of social organisations and civil society to improve peasant participation in development. I firmly believe that this should be done through the formation of peasant cooperatives because social organisation and agency are mutually linked phenomenon in the resettled areas (Murisa, 2009, p. 25). Scholarship should acknowledge that the biggest problem that undermines peasant participation is that the countryside is highly disorganized. The disjointed peasant voice is always ignored before, during and after the formulation or implementation of development programs. This is the reason why markets (sometimes with the help of the state), quickly acquires a hegemony in the countryside. "We need to buy fertilizers and chemicals from the market, also need to sale to the market to get money for cooking oil, sugar, salt and to send you to school", my grandfather would say lamenting to how market contact was unavoidable.

Linking the agrarian question to cooperatives

Lenin and Engels discussed the social and political dimension of the AQ, respectively, and then Chayanov added to the discussion, an economic dimension (Moyo et al., 2013, p. 104). Chayanov's conceptualisation of cooperatives is exceptionally appealing to the AQ because his definition of cooperatives has both entrepreneur and social movement characteristics (see page 57). Data to support this is then presented in later chapters.

The contemporary AQ has a variety of dimensions which may be separate but simultaneously converge and overlap with each other. These may include *among other things*, the land question (because of new forms of land alienation), the national question, the peasant question, gender question, ecological question, the agrarian finance question, the labour question, the industrialisation question, liberation, and the regional integration questions. The cooperative development model addresses some of these forms or dimensions (Figure 2.1). Cooperatives have a broad spectrum of peasant challenges they address, including the AQ of ecology and sustainability. Cooperatives are formed by the farmers whose primary concern also includes the need to preserve their land so that they can bequeath it to their offspring. This enables them to exploit their resources sustainably, take care of the environment and most importantly, profit from the utilisation of own surplus-labour, a benefit which is usually absent within a free market set-up. This was indeed grandfather's dream, for him to give his farm to my fathers who would then give the farm to me. Sustainability of the cooperative itself is achieved through the concept of self-help, which means they do not have to heavily rely on external financial sources once the cooperative kicks off.

Figure 2.1 is a summary of the AQ and the respective cooperative responses. By nature, cooperatives are best suited to fight the burgeoning level of poverty, inequalities, and social exclusion as they identify a variety of economic opportunities for their members. Individual risk is reduced through collective risk-taking. In Tanzania, multi-purpose cooperatives go further than Savings and Credit Cooperative Organisations (SACCO) which focus on the provision of finance (Makochekamwa, 2015). However, multi-purpose cooperatives transcend into other non-farm income projects (payment of cooperative dividends is a source of non-farm income itself).

Figure 2.1: Linkages between the agrarian question and the cooperative movement

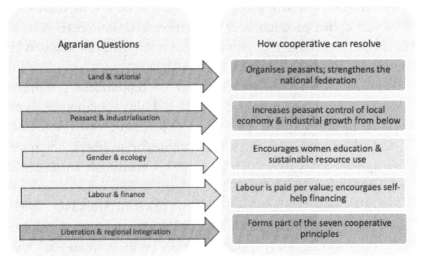

Source: Created by author

Perennial peasant problems such as remoteness, lack of access to information, poor infrastructure, low access to inputs/output markets and inadequate access to loans are solvable through the cooperative path. This is because cooperatives simultaneously undertake group marketing, credit mobilisation, information dissemination, progressive education, foster innovation, skills and capacity building for its members (ILO, 2015, p. 7). Within their community, these organisations can do infrastructural development which may be too expensive for the poor governments or may be unprofitable for the private sector.

There is evidence of high rates of women participation in cooperatives in countries such as Tanzania and Japan. Women's membership in some non-gender specific cooperatives[9] reaching as high as 65% and 95% respectively. Women also have gender-specific cooperatives to fight inequality and social exclusion and are thus geared towards resolving the AQ of gender. However, there still is much potential for the cooperatives to bring women further to the front line in agriculture through improving access to education. Inequalities in education affect access to information, skills level, land ownership and access to finance. This is why women's participation in cash crop cooperative groups was found to be low in studies of cocoa and coffee production in west Africa and some parts of east Africa (ILO, 2015, p. 7), or cotton, sugar and tobacco production in Zimbabwe (Mazwi & Muchetu, 2015, p. 27).

Scholarly debates on the best way to support the peasantry focused on how the market (capital) and the government were supposed to interact with this peasantry to ensure sustainable and equitable development. Some scholars argue for neoliberal doctrines – free agricultural markets (Badiane, 1997; Ortmann & King, 2007; Royer, 2014; Samuelson, 2012; Yamashita, 2015b); others focus on livelihoods and welfare economics (Akwabi-Ameyaw, 1997; Chayanov, 1991; Hayami & Godo, 2005; Kawagoe, 1999; Teruoka, 2008; Wiggins et al., 2011). Others are more on the extreme left – the radical political economy and the Marxist theories (Amin, 2012; Banaji, 1976; Bernstein, 2009; Egan, 1990; Jossa, 2014; Marx, 1973) that argue for the protection of the peasantry from market exploitation. The significant advantage of using theory in a scientific study is to simplify complex realities and be able to isolate specific components of a phenomenon and understand their relationships. We seek to simplify the challenges of the AQ through theory in this chapter.

Hayami Yujiro's Community-Market-State framework

Hayami Yujiro developed a Community-Market-State framework in his book *Development Economics: From the Poverty to the Wealth of Nations*. The framework was premised on the presence of market

---

[9] In Japan, more women are in consumer cooperatives than in agricultural producer cooperatives.

failures (caused mainly by information asymmetries) in a market-led economy (Otsuka & Kalirajan, 2010, pp. 3–4). Free markets alone cannot provide the necessary public goods (market failure), a government that is expected to provide public goods tends to be inefficient (government failure). In such a case, community institutions can address these failures through collective action from below. Hayami brought the community dimension to overcome the conundrums of state-market dualism. Information asymmetries are highest in developing countries because, in reality, different people know different things and reconciling their knowledge on the market is hugely challenging (Stiglitz, 2002, pp. 469–470).

In the rural areas, the state usually does not have enough information about the behaviour of the other players, i.e., the community and the corporates. In such instances, the CMS theory encourages the community to participate by providing information to the government and the market. Secondly, information asymmetries also exist between the corporates and the community, and when the community engages with the market, they do so without adequate information (i.e., prices, demand, supply, quality). This is another source of market failure; thus, it becomes the state's role to provide the community with information. The market also fails, and so does the state/government, hence the need for the community to chip in and remedy the situation (Hayami & Godo, 2005, pp. 246–248).

Scott's (1998, pp. 3–6) compelling description of how historical challenges encountered by the state in developing rural spaces solidify Hayami's argument. Historical failures by the government (collectivisation in Russia, communes in China and Ujamaa in Tanzania) were so severe because it imposed programs on the people and ignored local indigenous knowledge in designing programs or development interventions (Scott, 1998, p. 3). Often, state planning has neither the provision for indigenous technical-knowledge nor open improvising in the face of unpredictability. More often than not, the state is worried about improving its ability to collect taxes from the peasant surplus labour. Conceding that Scott's writing is described by some scholars as anarchist, including himself (Scott, 2012), I can still draw a couple of lessons from his argument.

A weak civil society is not desirable, especially in the presence of a high modernist state. Adopting Hayami's idea on developing countries like Zimbabwe is more appealing as indeed information asymmetries do exist (MAMID, 2012, p. 10; Samuel, 2018, p. 38; Stiglitz, 2002, p. 483). Hence, assimilating the community into the development agenda becomes salient. Recent efforts by development agents to incorporate community voices in development has escalated. However, these efforts neglect the fact that communities are unstructured, voiceless, and unorganised, especially in the case of post-FTLRP Zimbabwe. The reason is that socio-economic hardships have persisted in the rural space from the ESAP era and worsening in the post-2000 land reform (Makamure et al., 2001, p. 10). The challenge for the government of the day is figuring out the best combination that integrates the community-market-state efforts.

If a CMS framework is adopted, the community becomes a stakeholder of development and not just the passive subject of development (steps 2 in *Figure 2.2*).

Figure 2.2: Development of Hayami's CMS framework

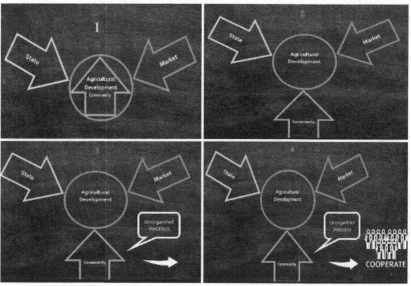

Source: Created by the author based on Scott (1998), Hayami & Godo (2005) and Otsuka & Kalirajan (2010)

However, community participation as a real stakeholder in development is hindered by the presence of disarticulated socio-organisational structures within the rural community; hence, the community has an unamplified voice (see steps 3 in *Figure 2.2*). This is mainly a problem in the context of post-reforms when disarticulation of social structures is highest. It takes time to cultivate social capital, construction of kith and kinship is a dawdling process. Solutions for a vulnerable community, a troubled government and an 'abusive' corporate market must be sustainable. Structural transformative efforts must link land reforms, agricultural financing, production processes, modern urban agro-industrial development and global community conduits for the benefit of vulnerable communities (Moyo, 2015, p. 38; Moyo & Yeros, 2005a, p. 23; Muchetu, 2019a, p. 38). In this sense, the establishment of structures of collective action amplifies farmer's voices. I then call into action the cooperatives model. The role of the cooperative is, therefore, to provide an organisational structure at the local community level that enables more direct amplification of the peasant voices. A unified peasantry will be able to reduce information asymmetries within the CMS framework.

Three main lessons from the above discussion apply to the case of Zimbabwe. Firstly, the marginal utility of importing inputs and technology is too high because the use of high-yielding inputs in Zimbabwe is still constrained. Although rapid growth in the use of hybrid seed and technology was realised from independence until ESAP era, the rate had significantly reduced after the land reform programme primarily due to low local production, for example, the fertiliser industry relied on imports in 2003/04 season (Rusike & Sukume, 2006, pp. 292–294). Thus, a simple improvement in the importation of new and efficient inputs will have a higher impact on the overall production. Secondly, just as in the above case, the FDI levels in the country have been at their lowest particularly after the land reform which saw government taking over most of the small-scale agricultural financing through subsidy loans (Moyo et al., 2014, pp. 17–18) funded by Iran, Egypt, Malaysia and China (Zumbika, 2006, p. 343). If government policy takes steps to correct this, the results in the agricultural sector can be phenomenal. Lastly, and more generally, the levels of agricultural education in the countryside are relatively low, which affects the spread of information, technology, and innovation. Improvements in these areas can go a long way in solving the challenges of the countryside.

Mapping the trajectory of the global cooperative movement

As discussed in the last section, the role of the state and the private sector in development is omnipresent; however, I argued that the part of the community is equally important. Here, it is necessary to define what a cooperative is before a more in-depth discussion on cooperative theories.

Cooperatives, what are they really?

Although many scholars use this term loosely to refer to a group of farmers organised and working together, a more in-depth analysis of the word may disqualify a few peasant organisations as cooperatives. Far left and far right scholars all have different definitions that try to conceptualise what cooperatives are. Emillianoff (1948) took us through the development of the meaning of cooperatives in both Marxist and neoclassical theory. He highlights that there are a few studies that have been able to give theoretical context to cooperative studies adequately. However, one of the notable attempts was by Tugan Baranovsky (1922) in his book *Social Basis for Cooperation* in which he argued that there are three types of cooperatives i) Proletariat cooperatives ii) a cooperation of the peasantry iii) cooperation of the urban middle class (petty bourgeoisie). Tugan Baranovsky (see page 57), whose definition of the cooperative movement formed the entrepreneurial side of Chayanov's (1991) broad definition, believed that the above three groups of cooperatives differed significantly in the character of their socio-economic organisation, economic purposes or goals pursued and most importantly, they differed in the cooperative ideas (Emelianoff, 1948, p. 17).

Valenti (1902) (cited in Emelianoff, 1948, p. 17) viewed cooperatives as intrinsically part of the existing global exchange system and was by no means trying to replace it. This rationale is in

30

line with some contemporary socialist that believed that cooperatives should not abruptly upsurge the capitalist model, but that capitalism was supposed to transform into a new model of cooperative production gradually (Jossa, 2014, p. 5). Cooperatives were supplementary institutions in these societies, and they assumed economic individualism. Although common problems are solved through collective efforts in a cooperative, it does not depreciate the individualistic nature of cooperatives. The fact that the unit of production in a cooperative remains as the individual household makes cooperatives superior to collectives (where the unit of production is the collective). Thus, a Valenti cooperative was "an economic system which, within the existing system of free competition, aims to wholly or partly correct the natural imperfections of the distribution of wealth." –Valenti (1902) (cited in Emelianoff, 1948, p. 19).

This view of cooperatives means they are not an entity that seeks to fight and oppose capitalistic institutions *per se*, but they are a corrective mechanism for the imperfections of capitalism and unequal distribution of wealth. In using this definition, not all cooperatives are cooperatives in the strict sense of this definition. All institutions that are not in the business of correcting some imperfections, inefficiency, unfairness, or some injustice in the distribution of wealth disqualify to be cooperatives, e.g., credit unions and insurance unions because they create and concentrate wealth rather than redistribute it. However, new credit unions and insurance unions facilitate access to inadequate resources for peasants otherwise excluded by formal finance institutions. Cooperatives do not necessarily compete with capitalistic entities, but instead, they come in to correct and supplement their deficiencies. This sets cooperatives apart from agricultural unions, trade unions, political parties, or any other social groupings (in terms of organisational objectives, ideals and economic purposes pursued).

The ICA has a different definition which has widely been used throughout the global movement. The ICA is discussed in greater detail on page 35.

31

The Rochdale Equitable Pioneers Society

Any discussion of the modern cooperative movement must begin in Rochdale town, England, where the first official formal cooperative – the Rochdale Equitable Pioneers Society – was born. A discussion of the agricultural cooperative movement must also peruse cooperatives in Germany, where the first agricultural cooperative – the Raiffeisen Cooperative Movement – began.

The founders of the Rochdale sought not only to establish fair prices for members of their consumer cooperative but also 'pure food', 'honesty', education, political rights and more equitable women participation in the local economic spheres of Rochdale town (Thomson, 1994). There was widespread unemployment in this weaving town with over 40% unemployment rates during the era of the cotton spinning mills and factories. The working, housing, and health conditions of the few employed were at their minimum. The pioneers aimed to create conditions necessary to end pauperisation and increase development. As I shall discuss, the pioneers were so successful that in the later stages of the cooperative's existence, it had to be forced by legislation to reduce their contribution to local education which was set at 10% of their profits annually (Thomson, 1994, p. xx).

The Pioneer Society was established based on the teachings on two early co-operators. Robert Owen (1771-1858) is believed to have been the first person to coin the term 'Cooperative society' in his magazine *The Economist*. He is also credited for coining the term 'Each for One and One for All' which has been used by several organisations across time and space, particularly in Japan where Owenism is said to have a more substantial presence (Thomson, 1994, p. 11). Dr William King (1786 – 1865) was, to a greater extent a scholar of Owenism. He managed to spread his ideas across Britain when he established a newspaper called *The Co-operator* in 1828. It is one of these newspapers that one of the early Pioneers (James Smithies) got hold of, studied, and fell in love with the idea of cooperation.

James Smithies introduced the teachings of Dr King and Owen to his fellow brothers. In 1930, a cooperative titled the *Rochdale friendly cooperative society* was formed based on a concoction of Owenism and

Dr King's ideas. They created a cooperative consumer store three years later, *the Rochdale Equitable Pioneers Society* located on number 15 Toad Lane in Rochdale. In addition to social reform, embedded in the Rochdale movement was the desire to improve the lives of their workers and members through enhanced access to health, education, and housing (Thomson, 1994, pp. 14–15). One of the reasons why this movement stood out and became influential throughout the cooperative movement is the fact that it had a deep understanding of the need for vertical rather than horizontal integration (cooperative rather than collectivisation). Horizontal integration meant achieving economies of scope by upstream cooperation, e.g., in the financing, farming, and manufacturing. This made it conform to the typical Chayanovian cooperative (discussed later, from page 52). Therefore, the Rochdalians managed to combine community with commerce, making sure that the resulting profits of this entity were equitably shared, and hence reduced the proliferation of the system of the haves and have-nots.

The Rochdale cooperative store only lasted a few years because it had overlooked the importance of reducing the number of bad debts; this is very critical for developing countries and their respective cooperative movements. Since it was a member-driven organisation, they had tried to charge lower prices for the consumers, hoping to cover for these low incomes through profits from the credit business. Also, the cooperative was not registered under the then existing Friendly Societies Act, and thus they had no legal standing to sue the bad debtors in the courts of law. However, the profits eventually could not cover the losses, and they had to liquidate their assets. Several other cooperatives that were created along the same lines of Owenism and King in England later suffered the same fate as the Rochdalian cooperative. Although all these movements later failed, they laid a foundation and provided lessons for future cooperatives.

The Raiffeisen movement

This cooperative movement, named after a parishioner by the name Friedrich Wilhelm Raiffeisen (1818-1888), was established to provide credit to German farmers who were struggling to raise capital for production. Unlike the Pioneers, this cooperative movement was located in the rural areas of Germany and fell under the producer cooperative, hence more applicable to the subject matter of this book. The cooperatives also came in as a solution for the failing rural agricultural markets in the mid-1800s. Raiffeisen, a community mayor at that time, used his influence and organised a group of wealthy community members to put their money into a single fund. This money would then be used to purchase food to feed people who were facing the devastating impacts of a famine in the 1846/47 season. However, this was not charity. The farmers received aid as a loan to which they had to repay when the famine was over. This aid society became the basis and a test run of his ideas about the community and how it had to restructure itself to fight famines, poverty, and underdevelopment (Klein, 2009).

Just as in the case of the Rochdale, the Raiffeisen cooperative movement was not the first cooperative movement in Germany (Viktor Aimé Huber had established housing cooperatives before), nor was it the only one at that time. Hermann Schulze-Delitzsch, a former judge, a member of the parliament and also a bourgeois started a relatively successful cooperative (credit and savings cooperatives) catering for the middle-class city workers (DGRV, 2009; Prinz, 2002, p. 12). Some scholars argue that the most successful of these movements was the Schulze-Delitzsch cooperatives as they led to the establishment of the most potent cooperative banking and credit institutions in Germany and beyond. In addition to establishing these credit and savings cooperatives, Schulze-Delitzsch also played a significant role in formulating and establishing the German cooperative law of 1881 from which the Japanese cooperative law was modelled on (Ishida, 2002a, 2003). A quick scan throughout cooperative literature also supports the success of these cooperatives measured through increased incomes to members, improved access to finance, markets and social services (education) otherwise unavailable (Emelianoff, 1948; Esham et al., 2012; Holyoake, 2016; Prinz, 2002; Schaars, 1971, p. 70). However, application of the Schulze-Delitzsch cooperative is broad, mostly catering for the worker and city dwellers, while the Raiffeisen cooperatives are more inclined towards the agricultural sector, with the farmer being the centre of the cooperatives. Inevitably, the

Raiffeisen cooperative structure seems to dominate the organisational structure of Japanese agricultural cooperatives. Accordingly, my book focused more on the Raiffeisen cooperative movement in Germany.

When these Raiffeisen cooperative ideas came to Japan, its society was much more homogenous, and farmers had more common grounds which enhanced their need and potential for cooperation (see Chapter Three). The case of Zimbabwe, on the other hand, presents itself as mottled, considering the FTLRP which redistributed land to over 180,000 families from various parts of the country (Moyo, 2011a). At times, beneficiaries from different ethnic groups or tribes were resettled next to each other, with diverse cultural values, language, and ways of doing things (Mkodzongi, 2016; Moyo et al., 2009; Murisa, 2009).

International Cooperative Alliance (ICA)

The ICA is the global cooperative body which seeks to unite, represent, and serve all matters relating to the development of one billion global cooperative members. It is one of the oldest non-governmental organisations, with over 310 national level organisations across 109 countries worldwide (ICA, 2019). It is a not-for-profit organisation that was established in the aftermath of the first Cooperative congress in August 1895 (in London) to advance further the role of the cooperative model in development and poverty reduction. It is currently headquartered in Brussels and has stood the test of two world wars, a cold war, and the current geopolitical shifts in the social lives of its members and liberal economic environment. Its principal focus is to provide a voice or platform for members to access knowledge, expertise and coordinated action to enable them to influence legislative environments to favour their sustainable development. The ICA has the mandate to guard the Statement of the Cooperative Identity (the ten cooperative values and seven cooperative principles).

To understand the development of this organisation, and how it united voices across different nations, it is worthwhile to appreciate its trajectory since its formation. By the year 1922, they had managed to set up the International Cooperative Banking Association (ICBA) and the International Cooperative and Mutual Insurance Federation (ICMIF). These organisations fundamentally became the foundation of the organisation as they were able to generate income for day to day operations (ICA, 2019).

In 1923, the first ICA day was commemorated on the first week of July. It is sad, however, that the first cooperative in the agricultural sector had to wait half a century later to be established (1951). Its primary purpose was to regroup agricultural cooperatives and was called the International Cooperative Agricultural Organisation (ICAO). This part of the ICA is of particular interest because it seeks to improve the economic, social, and cultural welfare of farmers across the globe. They intend to achieve this by enhancing the level of communication and cooperation within, beyond sectoral boundaries, and across nations.

In providing an efficient and compelling voice, the ICA continually refers to the Cooperative Statement, which contains a full description of the cooperative movement and a set of values which in turn informs the cooperative principles. Cooperative members worldwide agreed that the ICA should bear this task. The principles were formulated by firstly identifying a set of values:

- **Self-help**- people have the will and power to improve their lifestyles by working together as opposed to working in isolation as individuals.
- **Democracy-** members have the right to take an active part in decision making about issues that affect their lives through being heard, being informed, and participating.
- **Equality-** equal rights and opportunities for all persons who are members of the society.
- **Equity-** This refers to the need for equal distribution of resources.
- **Solidarity-** this describes the belief that there is strength in self-help and that members can overcome any challenges. Solidarity can also be established between cooperatives in the same sector, the same country and even across countries and continents.

• **Ethical issues**- these include the need to be extremely honest in doing cooperative business, to be open, to have a sense of social responsibility and to have a high level of caring for others.

The ethical values are highly regarded within Japanese society, and hence, it was easy to translate them into cooperative movement. The ICA produced a set of principles to help put the discussed values into practice. Principles are not merely a 'stale list to be reviewed periodically', but they are guidelines that show how to put ideas and values into practice (Hoyt, 1996). On the other hand, the cooperative principles were derived through an iterative process of co-operators' practical consensus at that present time. Hence principles are not static and should change with time (Birchall, 2005, pp. 46–47).

Principle 1: Voluntary and Open Membership

Co-operatives are voluntary organisations, open to all persons able to use their services and willing to accept the responsibilities of membership, without gender, social, racial, political, or religious discrimination. However, several cooperatives movements have violated this principle, especially in cooperatives that were formed based on local authority lines such as a village cooperative or a rural cooperative (Birchall, 2005). Most Cooperative society laws stipulate that a cooperative need to define its area of influence, which means a cooperative cannot have members from other places other than to which it is registered.

To some extent, a member is 'forced' to join a particular cooperative simply because it is the only one in their areas. In their study, Oczkowksi *et al.* (2013, p. 55) found that the level of adherence to principles differed according to the type of cooperative, and the reasons why the cooperative was formed. For example, in specific cooperatives like women cooperative, only women can join the cooperative, which means membership is not 'completely open'. The same scenario mimics the tobacco or cotton or sugarcane producer cooperative in which membership is attached to a specific product that is being produced. In this case, some principles may be violated and understandably so. Ishida's studies in the Japanese cooperative movement highlight how the principle of open membership and that

37

of autonomy and independence took less precedence in the movement as the focus was put on the stability of the movement which is threatened by *too much democracy and openness* (Ishida, 2002a, 2002b, 2002c, 2003).

Principle 2: Democratic Member Control

The predecessor of this principle read 'one member one vote' and suggested that every member was equal in terms of decision making and directing the trajectory of the cooperative. However, this does not apply to secondary cooperatives where small-sized and larger sized (in terms of membership) – cooperatives formed its membership. Thus, proportional representative voting is used to achieve democratic member control based on the size of the cooperative (see Hoyt, 1996).

Principle 3: Member Economic Participation

Since 1996, cooperatives are now enjoined to compensate capital and labour fairly. Members can profit more from participating in cooperative activities (Hoyt, 1996). Increasingly cooperatives have been open to sourcing capital from non-members and other outside financial institutions. Out-sourcing threatens the member economic participation principle as there is the risk of losing control of the cooperative and eventually losing the ability to profit from the activities of the cooperative for the members. That has been one of the most significant challenges for developed cooperatives such as the JA as the amount of trust has reduced such that the members feel that the number of benefits they receive is not even half of what they are supposed to receive (Ishida, 2003).

Principle 4: Autonomy and Independence

Cooperatives must be free from government and any other interferences. That significantly affected third world country cooperatives in which initially, state intervention was necessary to set up the cooperatives. This principle has not been practised as several governments especially in developing economies have been slow in relinquishing power to the cooperative structures; additionally, cooperatives are still seen as an extended arm of the state for channelling development and policy programs. Another way loss of independence and autonomy has been through finance capital. With the rising globalisation rates, increasingly cooperatives in both developed and developing countries, have accessed finance or debt from outsiders. That has directly compromised their level of autonomy.

Principle 5: Education, Training, and Information

This, for me, is the most important principle for developing countries. This involves teaching, educating, and training, not only the leaders, members, and young people who are still in school. If cooperatives are to be fully utilised as a part of the solution to agricultural problems, then people need not only know of the existence of the cooperative concept but should have a good appreciation of the movement and be willing to take part in it (Hoyt, 1996). Governments and national cooperative structures should undertake this task with the utmost urgency. Government and cooperatives have applied this principle to varying degrees depending on its training, education, or information dissemination.

Education and training have been limited to the leaders and the administrative echelons in developed cooperatives, while in developing countries, there is a shortage of educated leaders who can take the movement forward (ILO, 2015, pp. 7–9). On the other hand, information dissemination has generally improved with improvement in technology (Birchall, 2005, p. 56). Many educational courses/programs hardly have modules that focus or even mention cooperatives, and scholarship has not chipped in to help in this regard. Few scholars are writing about cooperatives' role in development, which has led to a continued ignorance in the way cooperatives are perceived by the general people, policymakers, and development agencies (ILO, 2015, p. 16; Pinto, 2009, p. 2). National federations should take this task.

Principle 6: Co-operation among Co-operatives

This principle has not changed since 1966; it encourages the working together of cooperatives across different sectors of the economy. It would not make sense for cooperatives to not work with each other when their sole aim is to unite people in building a better future. Most cooperatives have been following this principle. In the Global North, local cooperation has been achieved, but the cooperation has been reluctant to go beyond national borders and partner with developing cooperatives in emerging countries (Birchall, 2005, p. 58). There is a need to highlight the importance of the alliance between cooperatives in different borders as one may produce and export, then the other may import and process and sale to its members.

Principle 7: Concern for Community

It stresses that the cooperative has a social responsibility to preserve and efficiently use natural resources in a manner that allows generational sustainability. This principle also highlights the fact that if cooperatives took care of themselves as a group, then they would be able to lift themselves out of poverty. Indeed, working together would result in a better economic welfare state than the current one as shown in the developed world (e.g. Japan) where cooperatives developed their communities (Birchall, 2005, p. 58). However, this was achieved to limited levels in developing countries such as Zimbabwe, Zambia and Malawi (Otsuka & Kalirajan, 2010), due to low capability issues.

While some scholars have used globalisation to argue for the non-viability of cooperatives (M. Cook, 2018; Ortmann & King, 2007), others have used it to say for more cooperative approaches to be adopted (Amin, 2018; Jossa, 2018). This book belongs to the latter and believes that large scale cooperatives such as the JA in Japan and the Raiffeisen in Germany should simultaneously strike alliances with developing ones in the Global South while serving their local members. Conversely, smaller cooperatives such as those in the developing countries should act locally and think globally by kick-starting local industries.

The chapter has so far described the theoretical underpinning for including the community in agricultural development and how cooperative organisations achieves this. I have also given a tentative definition of cooperatives and the origins of the cooperative movement. In the sub-section that follows, I want to examine how cooperatives are understood through different theoretical frameworks to choose an ideal theoretical framework to follow throughout the book. I begin with the neoclassical theories, then move to the Marxist theories before discussing Chayanovian ideas on cooperation which I later adapted for my analysis.

The inappropriateness of market-based cooperative theories

Neoclassical economics is a social science that gives ultimate power to the market to decide the welfare of the people through the automatic market price determination mechanism. It postulates that if demand and supply (affected by the cost of production, taste, and preferences of the consumer) of a good or service are at equilibrium, market efficiency is achieved (Ortmann & King, 2007; Royer, 2014). It assumes that the firm's challenge is to maximise profit and that the need to accumulate surplus drives all types of organisations in an economy. This line of thinking, therefore, makes it extremely difficult to apply this theory to cooperatives which are not profit-maximizing. Historically, not only did the cooperative principles avoid focusing on maximisation of profit, but they discouraged it, primarily if it occurred at the expense of other members' livelihoods. Cooperatives are mainly concerned with serving their members often at cost price, and hence neoclassical theory has struggled to find the nexus between business theory and cooperative practice.

The key questions asked is if an organisation does not principally want to make money (surplus value), then why is it in the market? What motivates this firm? How can it be financed? Moreover, is it sustainable in a capitalist society? However, with the advancement of the capitalist system and the realisation of the various contradictions that it brought, neoclassical scholars are beginning to realize the potential role that cooperatives could play in development (Egan, 1990, p. 73).

Neo-Classical Economics and cooperatives

Iliopoulos (2017) divides the development of cooperatives in the 20th century into five distinct stages depending on the approach used, the focus or the goals of the cooperatives, the tools used, and the

41

assumptions made about the movement (see *Figure 2.3*). The first stage, market-oriented scholars in this era viewed the agricultural cooperative movement as an extension of the farm, and the overall objective was to optimize farm operations and productivity. Secondly, the cooperative was considered as a separate entity from the farm and hence was supposed to become multi-purpose through increased production and vertical integration. In this stage, the cooperative is more engaged in the market and acts like a corporation (firm) and is involved in the inter-cooperative competition. Many of the cooperatives are still viewed in this way until today (Royer, 2014).

Figure 2.3: Evolution of neo-classical cooperative theory in the 20th century

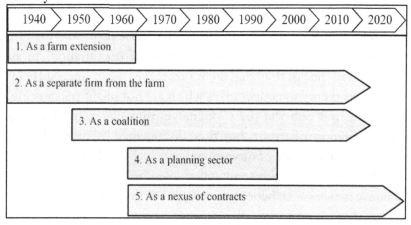

Source: Created by the author based on Iliopoulos (2017, p. 4)

The third stage perceives cooperative as a coalition for improving management decisions and organisational structures at the farm level; additionally, they focus on intra-cooperative bargaining as opposed to competition. The fourth stage saw cooperatives as institutions that can help or improve the planning sector in a rural or national economy (*Figure 2.3*). They were concerned about the improvement of coordination within the movement. In the last stage, in which the majority of the classical scholars are found (Birchall, 2005; M. Cook et al., 2009; Royer, 2014), regarded cooperatives as a network of contracts whose primary goals are to improve economic efficiency and organisational design (Iliopoulos, 2017). The cooperative movement is advised to focus on property rights, improve governance and coordination, and encourage the state to formulate sound public policies.

The above illustrates how neoclassical (NCE) theories have struggled to describe cooperatives which led to a new breed of New Institutional Economics (NIE) theories found in the last stage. These mainly utilise games theory, the transaction cost theory, property rights theory and the agency theory to model cooperative behaviour (Ortmann & King, 2007). Although both NCE and NIE theories believe in the power of the market, the latter believes that the market if not perfect because information asymmetries exist, hence there is no ideal competition (Samuelson, 2012). And thus, justifying cooperative existence. In explaining cooperative behaviour, NIE theories focus on the completeness and incompleteness of contracts.

New Institutional Economics and cooperatives

To explain cooperative socio-economic behaviour from start-up, growth, and development, Harte (1997 cited in Royer, 1999, pp. 58–59) and Cook (2018; 2009) developed a conceptual Life-Cycle Framework (LCF). Harte's framework rationalised cooperative formation as a corrective measure in the face of free-market failures; however, when the correction has been done, the cooperative must convert into an investor-owned-firm. On the other hand, Cook's LCF provides more options depending on the type of cooperative, the socio-economic environment, and membership structure.

Building on a host of other earlier scholars who proposed as many as ten stages in a cooperative lifespan, five distinct stages that a cooperative goes through were identified (Figure 2.4). These were P1) justification stage – everyone realising the need for a cooperative and joining. Then P2) organising and designing the rules and principles of the cooperative followed by P3) growth, glory & heterogeneity – growth is realised until the heterogeneity issues begin to develop. Then comes the P4) recognition and introspection – is when the cooperative must look at where they are, where they have come from and where they are going. The last stage is the P5) choice – must be made either to reinvent or remain on the same path or dismantle the cooperative.

Figure 2.4: Cook's lifecycle of a cooperative

Health of Cooperative

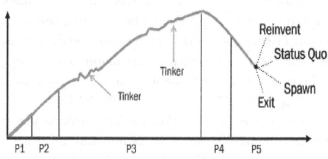

Source: (Cook, 2018, p. 7)

Neo-liberal theories advocate for the cooperative to reinvent into a private corporation (see more in M. Cook et al., 2009, pp. 1–13; Royer, 1999, pp. 57–59)

Given that markets are inefficient, and the potential for the opportunistic appropriation of quasi-rents from farmers in the markets, NIE argue that cooperatives may be necessary to maximise benefits to the members, to maximise the per-unit value or average price and increase the patronage earnings to the members. The cooperative must organise its institutional structures, make collective decisions, and respect property rights, just like an ordinary firm would do in a perfect market economy (Ortmann & King, 2007, p. 52; Royer, 1999, pp. 45–52). Thus, the more a cooperative behaves like a private firm, the more it can be explained through NCE and NIE theories.

The **Transaction Costs Theory** refer to the costs of organising and transacting, which occurs in every part of an exchange process. The process involves wide-ranging costs, including those associated with soliciting information, bargaining, decisions making, policing and enforcement, which all occur within a political, social, economic, and legal environment (Ortmann & King, 2007, p. 54; Royer, 1999, p. 46, 2014). More market participants prefer to act in a market with adequately freely accessible information, and hence would instead enter into contracts to reduce the likelihood of information asymmetries. Information asymmetries lead to higher transaction costs and the exploitation of the farmers.

Furthermore, if a group of people working together as one unit can select institutional arrangements that reduce these transactions costs and correct market inefficiencies, this establishes the rationale for the existence of a cooperative.

The **Agency Theory** seeks to understand the relationship between an individual or group of individuals (agent) who acts on behalf of another organisation (principal). The objectives of management may become divorced from those of the capital owners, and hence, resulting in *principal and agent* problems. The theory argues that the centre of control within a cooperative business is ambiguous (Royer, 1999, p. 50, 2014). In an investor-owned firm, the investor, who owns the capital makes decisions on capital outlay, labour, and allocation of other resources which generates profits. Although cooperatives are said to be democratically member-controlled, the fact that a group of members will have varying goals and objectives makes it difficult to understand how they can agree on which combination of land, labour, and capital to use.

The management committee (agent) is chosen to run the cooperative on behalf of the peasants/community (principal). Very often, the management and supervisory committees must make decisions without the immediate approval of the members, which may lead to principal-agent problems. The principal-agent problem decreases member participation as the distance between principal objectives and agent principles increases. The absence of a defined centre of power makes it hard for market-oriented theories to model cooperative behaviour (Royer, 2014, p. 3). Just as suggested in the transactions cost theory, a contract (that bind the agent to act according to the principal's specific interest) should be established to reduce the principal-agent problems. Transactions costs focus on the transactions, while agency theory focuses on the individual or organisation (Royer, 1999, p. 50).

The **property rights theory** hinges on one of the underlying market-oriented assumptions which say the property is privately held and that private-property rights are far more superior than any other form of property rights. The theory argues that the nature of ownership of a set of assets affects its allocation and organisation during value creation (M. Cook et al., 2009; Ortmann & King, 2007). Conspicuously, this implies that any property rights structure that deviates from private property results in the inefficient reallocation of resources (Royer, 1999, p. 51). The theory further argues that the transferability of one asset to the next is extremely important for markets to work. Hence, compared to privately held firms, publicly held firms have less motivated stakeholders to supervise the management. Just as in the case for the transaction costs theory, contracts are also incomplete (because of information asymmetry). However, if property rights are individually owned, it means each party can use their private property rights to negotiate or bargain on how the asset's market activities should be carried out. Thus, the incompleteness of contracts necessitates the definition of property rights and necessitates assigning private property rights.

The cooperative movement is therefore defined, through this theory, as an association of assets under joint ownership. To improve the allocation and restructuring of that asset to create wealth, the cooperative must have an internal mechanism that clearly defines the property rights of those assets (Ortmann & King, 2007; Royer, 1999, p. 49). The property rights theory seems to argue for a cooperative ahead of collectives because the central unit of ownership of an asset or a line of production remains as the individual farmer for cooperatives. The farmer still has the bargaining power to negotiate for more efficient allocation of his/her resources to produce high value than in the latter.

How market-based theories problematizes the cooperative.

The theoretical analysis above-identified five primary problems in cooperative development; the free-rider, horizon, control, portfolio, and influence-cost situation. The existence of these problems forms the basis for market-oriented theory's argument to restructure and reorganise a cooperative towards an investor type of organisation as a solution as observed in the Japanese Cooperative movement. This forms the basis for the degeneration theory, which argues that eventually cooperatives risk converting into investor-owned firms.

The free-rider problem emanates from the property rights theory because individuals cannot face the full cost or benefit of their decisions or activities in the market place (Cotterill, 1988, p. 227; Hayami & Godo, 2005, pp. 13–15). Free-rider problems can occur internally, e.g., when new members get equal benefits as those of old members who went through the complicated initial start-up process. It can also happen externally, e.g. when non-members enjoy market conditions brought about by the efforts of the cooperative (FAO, 1998).

Horizon problems are based on the concept of poorly defined property rights as well. This problem emerges because cooperatives are concerned about satisfying their members in the short run. Members are said to underinvest into the cooperative depending on the intended membership period. That makes it harder for the cooperative to invest in research and education or other long-term investments.

Portfolio problems, on the other hand, arise because cooperative capital is based on equity shares with restricted tradability, which undercuts their ability to expand their portfolio. Members cannot spread their risk outside the scope of their cooperative while at the same time, those investors outside cannot absorb the risk because membership is restricted to the content of the cooperative (M. Cook, 2018, p. 10).

Control problems are derived from the agency theory and refer to the situation in which the principal and agent's objectives start to diverge (principle-agent issues). According to this theory, as compared to investor-owned firms, the absence of a market for equity shares in cooperativism means the principal cannot supervise and closely monitor the agent (Royer, 1999, pp. 55–56).

Finally, since the one member one vote principle decides cooperative decisions and the general direction of the cooperative, *influence costs* arise when members of the cooperative try to persuade other members to vote in a certain way. These costs can become too high and in worst cases, can become more like bribes. The theory highlights that although influence costs are found even in non-cooperative organisations with a more homogenous money-making objective, they are mostly higher for cooperatives because they have

47

more stockholders with diverse goals and objectives (Ortmann & King, 2007, p. 59).

Neo-classical and new institutional theories only understand cooperative if they assume them to behave like investor-owned firms. Although the theory enables managers to lead their cooperatives more professionally and hence improve profits, its conceptualisation of the cooperative problem – as income maximising – is weak. This is so because the cooperative seeks, simultaneously, social, political, and economic objectives which cannot be modelled solely through increased cooperative income. Therefore, I could not adopt this theory for my book.

Why Marxist cooperatives theories are weak.

In Marxism, what structures, principles and laws are cooperatives supposed to obey and what goals and objectives are they supposed to achieve?

Marx and cooperatives

For a long time, there had existed a misreading of Marx, and Marxist position on labour managed firms, as some scholars argued that Marx did not like cooperatives (Jossa, 2013). That was primarily based on the fact that Marxists regarded peasants as a potential capitalist producer whose contribution to the economy would not lead to socialism. Thus, Marx is said to have rejected cooperatives. However, it is important to note that Marxist scholars not only endorsed cooperatives as a new production model but also regretted not having focused on them much earlier. Although Marx's primary focus was not on the peasantry, his worker/labour-managed firms (proletariat) and the joint-stock company cooperatives formed by the petty bourgeoisie could be applied easily to the peasant cooperatives (Egan, 1990, p. 76; Jossa, 2018, p. 154).

Much of Marx's works also focused on class struggle, that everything is an endless class struggle between labour and capital. Peasant class struggle would result in capitalistic production; hence the peasant was inherently capitalistic. In this respect, cooperative offered a far more superior production model because it would result in a worker-owned firm in which the workers become their capitalists, exploit their labour, and valorise it at the same time. Cooperative's ability to differentiate the two streams of income from capital and income from labour makes it superior in reversing the imbalances brought about by the capitalistic peasant producer (Jossa, 2014). In doing so, cooperatives would resolve class struggle since it is easier for labour to work for firms owned by themselves (Egan, 1990, p. 73). Therefore, cooperatives were a means to an end and a way of achieving socialism. It would seem that Marx would use the cooperative to achieve socialism with less emphasis on giving control of the rural economy to the peasantry itself.

Lenin and cooperatives

Vladimir Lenin's article that was eventually published in May 1923 represents a significant turning point in Marxist thinking. The cooperative envisaged by Karl Marx took the form of a state-controlled institution, governed by workers or labour as is required under a socialist model. However, as a follower of Marx, Lenin soon realised the need to reduce economic nationalisation or central planning by giving peasants more control of their local economies. This was influenced by the formulation of the New Economic Plan in USSR, which Lenin thought of as a temporary retreat from socialism (Jossa, 2014, p. 4). For Marx, cooperatives were a transition from capitalism to socialism, or a means to an end and not the end itself. However, enter Lenin, and cooperatives had to be owned by the community itself, and were capable of jumping stages straight into socialism. Lenin (1923, p. 1803) explained that this was the best way of "setting our feet on the soils of socialism" and equated cooperativism to socialism.

Lenin goes on to argue that increasingly, economic power and control should go to the cooperatives instead of the state since the state and central planning was "deplorable, not to say wretched, that we must first think very carefully how to combat its defects" (Lenin, 1923, p. 487). Thus, there is a movement away from the advocacy of a state-owned firm exploiting hired labour, to a system of labour-controlled firms exploiting 'self-labour'. Lenin's article on cooperation managed to do two main things, *i)* to advocate for the

49

unification of the peasants and the proletariat to undo the hegemony of the large and private capital, and *ii)* it described cooperatives as socialist firms controlled by the working class on behalf of the state (Jossa, 2014). This line of thought supports the basis for this book.

Contemporary Marxism and cooperative degeneration

Historically, governments have pursued political democracy at the expense of economic and social democracy. In a political democracy, a country is desirably ruled by the people, and hence decisions made about the political welfare of the country are representative of the people's wishes (Jossa, 2014, p. 4, 2018, p. 8). However, when it comes to the economy and about the utilisation of resources, a democratic process is believed to be inefficient, and hence only a few capital-owners have the right to decide. Economic democracy can be achieved through the use of labour-managed firms (LMF) (Jossa, 2008, pp. 5–9; Winn, 2013). Workers make decisions about the firm, hire capital, and pay it at a specified interest rate and retain the balance.

There seems to be a silent consensus that classical Marxist cooperatives cannot be envisaged within an advanced capitalist model. Egan (1990, pp. 74–75) argues that most of the scholars associate Marxist scholarship to the *degeneration thesis*. The degeneration thesis says that cooperatives start as labour-managed and labour-controlled organisations, but then they grow larger to such an extent that they need to hire labour outside their cooperative (Reich & Devine, 1981 cited in Egan, 1990, p. 74). With further growth, control becomes wholly vested in the hands of the management, which eventually results in labour exploitation in the same way a corporate firm would.

Since cooperatives operate in a capitalist market system, it must act according to the rules of the game, i.e., like a capitalist firm. Eventually, it will try to increase the productivity of labour (as is required to survive in a capitalist set-up) by either self-exploitation of its labour or by hiring non-cooperative-member labour. And lastly, it is forced to compete with other cooperatives in the same market. Furthermore, the success of some cooperatives will result in the accumulation of wealth in one region which ideologically undermines the original concept of cooperatives. The hegemony of the capitalist framework is therefore seen as so overwhelming that it forces cooperative to behave like any other capitalist organisation. That is the rationale of the degeneration thesis.

This stage of a cooperative development equates to the LCF stage 4 (P5, see page 43) in which the cooperatives have to decide whether to tinker, reinvent or exit. However, to counter this problem, the contemporary Marxist cooperative theory then stipulates that there should be a national federation (Marx, 1973). This federation should oversee redistribution and governing of issues or contradictions that may arise between the cooperative movement and the market (Egan, 1990, p. 79; Markowski & Vanek, 1972). The federation should reduce class struggle, to ensure intra- and inter-cooperative *solidaristic orientation*. This, together with a deliberate but moderate connection with political struggle, makes cooperatives unite and help each other instead of competing.

Egan (1990), however, argued that cooperatives could overcome degeneration. In the beginning, there is a capitalist who is directly in control of the capital. As the firm grows, the capitalist will hire management to look after capital and to employ labour. The capitalist no longer needs to carry out all the functions, and this creates a joint-stock firm in which management (who are also workers) are in charge of labour instead of capital. Thus, at this stage, Marx argued that social production and private property are now divorced from each other, but since this divorce occurs within the capitalist production framework, the *resolution is negative* (Marx, 1992, pp. 571–572). A negative resolution of the capital-labour problem means labour is still sold to capital, which continues to exploit it, this time through the management. In the case of cooperatives, the contradiction between management and the labour is removed since labour hires management and not the other way around.

Thus, three issues were suggested to avoid degeneration in Marxist cooperatives; *i)* all workers must share in the firm to reduce the risk of recreating classes within the cooperative, this mostly means little to no hiring of labour from outside the cooperative, *ii)* all cooperatives must belong to a national federation which should be in charge of fostering common bonds and common grounds, and *iii)* a proportion of the surplus should be devoted to the creation of more capital and the creation of more cooperatives (Marx, 1973). Considerably, these remedies are reflected in the 1995 ICA cooperative values of ˜self-help, equality, equity, democracy, and solidarity. The relationship between the degeneration thesis and the cooperative is tied to the later commentary on Marxist cooperatives than the Marxist theorisation of cooperatives itself (Egan, 1990, pp. 77–81).

Introduction to Alexander Chayanov; A third-way approach

This theory was put forward by a Russian scholar, Alexander Vasilevich Chayanov (1888-1937) was a Russian scholar credited for the theory of peasant cooperatives in the late 1920s. However, it was only a half a century later (1987) that his works were recognised mainly by Theodor Shanin (Chayanov, 1991; Shanin, 2009). His studies mainly focused on the agricultural cooperatives in rural Belgium, Italy, France, Switzerland, and Russia in the 1920s. To articulate ideas on the progressive path that peasant societies should take, Chayanov first started with a working definition of the peasantry. His conceptualisation of the peasantry (page 54) enabled him to reduce the problem of the countryside to a reasonable level. However, his interpretation of the peasantry would also become the focus of most of his criticism, especially within Marxist scholarship (Banaji, 1976; Bernstein, 2009, pp. 56–59). Chayanov's model comes in as a third-way type and most useful framework for understanding present-day cooperatives in Japan and Zimbabwe. In this respect, I describe the background and context to Russia when Chayanov was carrying out his theorisation to succinctly explain how relevant his ideas are for the new Zimbabwean cooperatives.

A general agreement that cooperatives were the best form of social organisations that would deliver the peasants out of pauperisation swept across Russia as seen by an increase in the number of cooperatives from 1,625 in 1905 to 35,200 in 1915. These types of cooperatives were not bottom-up but were driven by either the scholarly works of the intellectuals or the state (Chayanov, 1991) or both. The general idea was to use cooperatives to achieve socialism, and this had to depend on the political conscientization of the peasants (as espoused in the Marxist-Leninist framework). Before Chayanov took centre stage, two significant scholars were writing about agricultural cooperatives in Russia. The first one was S.N Prokopovich (1913), who believed that there existed administrative and legal impediments to the cooperative movement. That individual rights (autonomy and freedom of association) had to be established before the cooperative could move forward (Chayanov, 1991, pp. xv–xvi). The second intellectual, Tugan-Baranovskii (1913) believed that although cooperatives were unable to achieve a shift from socialism to capitalism, they played a critical role in aggregating the peasants' voices to guard their interest. For Tugan-Baranovskii, the cooperative had the potential to build a new society within an already existing one.

Chayanov's pamphlet titled '*What is the agrarian question*' published by the League of Agrarian Reformers in 1917 made him unpopular with the ruling elites (Chayanov, 1991, p. xxvi, 2018). This was so because 'land to the working people' was central to his *revolutionary demand* (virtually a demand for the landlords to cede their land to the peasantry). He despised capitalism, but also did not outrightly support state socialism or anarchistic communism, but instead believed that a middle ground could be established (Chayanov, 1991, p. xxvii). His solutions to the AQ are termed 'third-way type' of interventions for this reason. So only cooperation could help establish this third-way common ground just as argued by Tugan-Baranovskii:

> A cooperative emerges fully equipped with capitalist technology, and it stands on the capitalist ground, and this is what distinguishes it in principle from socialist communes which seek to create an economic organisation on an entirely new economic basis. – Tugan-Baranovskii cited in Chayanov (1991, p. xxvii)

The revolution of 1916-17 placed Lenin's Bolsheviks into power, and they implemented a radical land reform by dictatorial methods (Chayanov, 1991, p. xxix). Thus, although Chayanov continued to work with the Bolshevik government, his antagonism to the October revolution was a public secret which set a series of persecutions under Stalin and culminated in his execution in 1937.

Understanding the peasantry according to Chayanov.

Chayanov made some daring inferences about the peasantry arguing that peasant farmer's production was motivated by use value rather than the exchange value. That is, small peasant farmers are primarily concerned with utilising their labour to a level consistent with household food security (Chayanov, 1991, p. xxv). The focus on labour was backed by what he called the labour-consumer balance concept. In this respect, peasant farmers utilised their labour on and off-farm to satisfy family needs, and only commodified their produce to meet the household needs that could not be met by their production, i.e., to buy such things as sugar, cooking oil and salt (the labour-consumption concept) (Thorner, 1965, p. 231; Thorner et al., 1986, p. xv). Chayanov's definition of the peasantry implicitly assumed the peasants to be rural based, with small-scale technological units and minimal capital-intensive technologies. The most exceptional observation, which influenced a number of his solutions and theories about the peasantry and peasant cooperatives was that the peasant farmers predominantly relied on family labour.

This was persistent in the countryside in both Russia and Asia in the first three-quarters of the 1900s and are currently pervasive in rural Africa. The peasant nature and condition determined the development and prosperity of cooperatives, according to Chayanov, in Russia and Asia; and hence should also support the development of these in rural Africa, I argue in this book.

Critics of Chayanov's theory of peasant cooperatives directed their criticism to his definition of the peasantry; this has profound effects on the theory of peasant cooperatives, which heavily relied on this definition. Harrison (1977, cited in Lyimo, 2012), a Marxist scholar, listed the flaws of Chayanov's ideas on the peasantry as follows; i) empirical data did not support the assumptions that low producer and consumer ratios caused low per-capita income; ii) he ignored the pervasiveness of peasant differentiation hence classified all peasants as not labour hiring and thus did not differentiate between rich peasants and poor peasants; iii) issues of specialization brought about by new technologies and population growth were not taken into consideration in Chayanov's definition of the peasantry.

Various Marxist critics of Chayanov form what has come to be called the Chayanov-Lenin debate. Indeed Bernstein's (2009) article on Lenin & Chayanov provides excellent insights into the differences between their theories of the peasantry. He explained that many of the critics of Chayanov believed that he opposed Lenin, but he opposed Stalin's policies instead. It made more sense to call it the Chayanov-Stalin debate instead (Bernstein, 2009, pp. 55–59). These two scholars had more points of convergences than divergence, and that Lenin had managed to receive more 'advertising' than had Chayanov. However, some scholars argue that the two differed on a couple of extremely fundamental issues, including the issue of peasant differentiation (Bernstein, 2009; S. Cook & Binford, 1986). For example, while Lenin emphasised on the presence of differentiation based on land sizes within the peasantry, Chayanov acknowledged differences in land sizes but underplayed its role in differentiation phenomenon.

Thus, Chayanov argued that land sizes were not the basis for differences in peasant development. After the 1921 land reforms, Lenin seemed to agree with Chayanov that peasant differentiation had decreased. That is particularly interesting for post-land reform cooperatives such as those in Zimbabwe. Chayanov and Lenin never actually debated or mentioned each other's names throughout their writings (Bernstein, 2009, p. 61). Ironically, some commentators claim Lenin wrote his 1923 article with insights from Chayanov's writings (Chayanov, 1991, p. xxxi). Despite criticism on Chayanov's theory of peasant cooperatives, it has remained one of the ideas grounded on the in-depth agricultural economic interpretation of data collected from the Russian peasantry and one dedicated to peasant cooperatives. It is worthwhile to pursue a deeper understanding of this theory. I strive to do this in the proceeding sections.

Development of the theory of peasant studies

In the early 20th century, Russia wanted to aggregate landholdings in rural areas to pave the way for large-scale production. In arguing against such a model of production, Chayanov used the analogy of the industrial and manufacturing sector in pre-capitalist Russia, which were once family-owned before the explosion of industrial capitalism. He warned that agriculture could also see the same fate, but horizontal integration in the agricultural sector would be a disaster (Chayanov, 1925). Furthermore, capitalism would penetrate through vertical concentration by controlling the channels utilised by farmers in accessing their input and output markets. By doing this, the capital then effectively gains control of the market by developing a system of credit and finance based on conditions amounting to slavery and hence turned agriculture into a network of exploitation of the farmers. This straightforward narrative was pervasive in the Asian agrarian sector, with particular mention of Japan where the feudal system had laid the ground for the development of capitalism; and the later landlord-tenant system merely continuing with an already established system of exploitation (see Teruoka, 2008).

African agricultural-based economies engaged in the production of commodities for export (mostly in their raw or near-raw form) are still subjected to these exploitative tendencies of the market. Production processes are dictated through the imposition of standards, quotas, planting dates and rotation schemes which effectively reduces producers to mere technical labour providers, a phenomenon that has been argued by some scholars as 'disguised workers'[10] (Little & Watts, 1994; Toendepi, 2018, pp. 31–32). The prevalence of contract farming for tobacco, cotton and sugarcane in the Zimbabwean agricultural sector is a typical example. Chayanov also argued that financial capital had other means of catching up with farmers and exploiting them as well. These included the mortgaging, credit, transportation, and irrigation systems which abetted the turning of the rural area into a source of labour, with no control of the means of production.

---

[10] Disguised workers ('concealed wage labour' or 'disguised proletariat') are usually landowner farmers under contract farming. The financing firm provides all the factors of production from inputs to extension (planting, harvesting schedules) and supervision such that the landowner virtually becomes a laborer on their own farm (Little & Watts, 1994).

It was this agrarian problem presented by the capitalist mode of production that resulted in the suppositions of rural cooperatives. Thus, if capitalist control of the means of production were to manifest itself as cooperative 'capitalism', which is controlled by the peasant, then small producers would not surrender their produce (and hence control) to the markets. They would engage with it collectively as associations of numerous farmers. The ordinary farmer is saved from the jaws of the capitalist dragon and is then able to connect to millions of other producers and consumers directly by selling through a cooperative (Holyoake, 2016). This manifestation of cooperativism can and should initially be supported by the state only through the legislature and state credit for the cooperative to evolve into a reputable organisation that eventually avails credit resources to the rural areas on its own (Hayami & Godo, 2005). Thus, regaining control of the rural economy, bringing vertical concentration, where cooperatives and the farmers are elevated to a better status in the economy and society (household); as farmers reconfigure their households to match cooperative goals. This conceptualisation seems to dispute Harrison's (1977) assertion that Chayanov's theory did not account for changes in household characteristics.

Defining a Chayanov cooperative

What then constitutes a cooperative, and what makes it different from corporations and capitalist farming enterprises? There is no one agreed definition of what a cooperative is; however, in developing the theory, Chayanov utilised two famous definitions of cooperatives:

> A cooperative is an economic enterprise made up of several voluntary associated individuals whose main aim is **not to obtain the maximum profit** from capital outlay **but to increase the income** derived from the work of its members, or to **reduce the latter's expenditure** by means of **common economic management.** (bold by the author for emphasis). –M Tugan Baranovskii cited in Chayanov (1991, p. 14)

And a second definition by Pazhitnov.

A cooperative is a voluntary association of some individuals which aims, by its joint efforts, to **combat the exploitation** (by capital) and **to improve the position of its members** through the production, exchange, and distribution of economic benefits, thus as producers, consumers, or sellers of labour. (bold by the author for emphasis) – K Pazhitnov cited in Chayanov (1991, p. 14)

From these two definitions, cooperatives appear either to establish the efficient organisational allocation of capital and distribution of gains, or those that focus on fixing socioeconomic (sometimes political) ills (class harmony and liberation of the peasantry from economic exploitation), or sometimes both. Additionally, cooperatives do try to improve the position of the members through increasing their incomes (making it appeal to market-oriented theory). Chayanov combined the two definitions to describe a cooperative with entrepreneurship and social movement characteristics. For that reason, Chayanov concludes that in coming up with a theory of cooperatives, it is necessary to define the 'cooperative' as representing a variation of the peasant economy with enablers for a small-scale farmer to enjoy 'large-scale benefits' without losing the individuality or control of his production process.

To further understand the significance of a cooperative and which type of cooperative in particular, Chayanov used an example of a butter processing plant that can be established in a milk-producing rural area or region. The butter factory can thus be owned by four distinct types of owners, as illustrated in Figure 2.5. The example holds for agrarian societies that have the majority of people living in rural areas such as those in African countries.

Suppose a producer cooperative (farmers) owned the factory and did production with the help of hired labour, then the benefits or 'profits' of this factory are shared among the cooperative members. Thus, the interest of the producer cooperative is to increase the price of the milk purchased from the farmers, which is desirable assuming that the motive is to develop the peasantry who make up the majority of the population (Stack & Sukume, 2006, p. 557). The second case is if that factory is under the control of a consumer cooperative which naturally aims to get butter to the consumer at the lowest price possible. What this means is that the farmer producer surplus labour is exploited and hence tend to lose out. The third scenario is if a worker's union/cooperative owns the factory, then the will and interest of the workers guide the cooperative. Thus, the workers buy the milk at a lower price from the farmers so that they can process and sell at a higher price to the consumers. The benefit from manufacture of the butter, in this case, is now only shared among the cooperative worker members. The farmer loses out.

Figure 2.5: Advantages of adopting producer cooperatives.

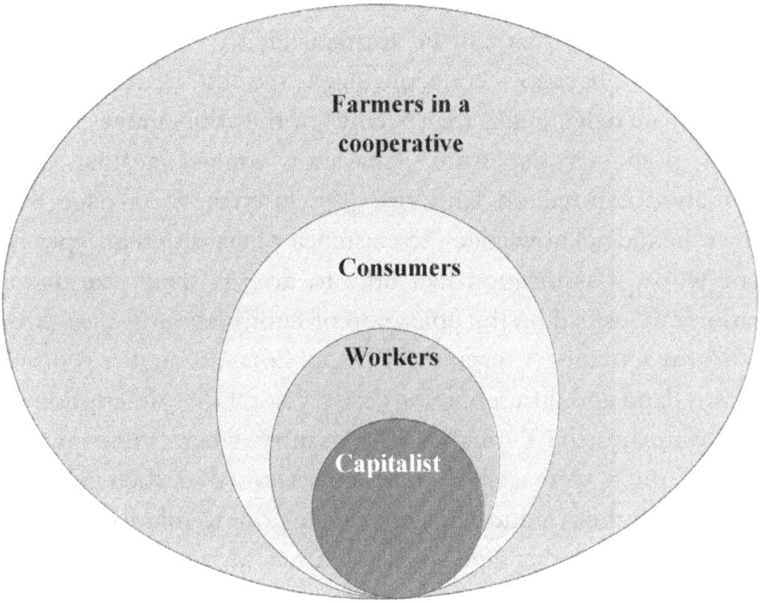

Source: Created by the author based on Chayanov (1991, pp. 18-20)

Lastly, the fourth scenario is where the factory is owned by one private entrepreneur whose primary goal is to increase his profits at all costs. This is the classic capitalist who will extract surplus labour from the farmers, the workers and still charge higher prices of the butter to the consumer to maximise his overall gain. The premise of the argument is that farmer (and their household) account for a higher proportion of the population in agrarian societies, and hence a system that benefits farmers also benefits a more substantial proportion of the population. The proportion of farming households is higher than consumer household (non-farmer consumers). Consumer households, on the other hand, overwhelm workers households who overwhelm the capitalist household. Thus, in an agrarian society, producer cooperatives engaged in vertical integration result in more equitable distribution of benefits/wealth.

The role and character of peasants in a Chayanovian cooperative

The theory has specific provisions or suggestions on the ideal type of cooperatives to be formed subject to the conditions in rural areas, nature of the peasantry and kind of crop/product under production. The nature and character of the peasants determine the type of cooperatives that can be formed. Hence, it is essential to describe how Chayanov conceptualized the differences in the peasants before delving into the specifics of the cooperatives.

As a point of departure, Chayanov argued against class differentiation centred on land size (see discussion on page 54). However, he did acknowledge the existence of six different types of farmers, whose classification had little to do with land size under cultivation but instead on the utilisation of family labour for on-farm and off-farm activities (Chayanov, 1991, p. 26). Although a positive correlation (land and differentiation) exist, Chayanov's description of peasants estimates the Zimbabwean peasantry, where some smaller land-sized farmers were well off than larger land-sized. Rich peasants can be found in the CA and A1 sector while poor to middle peasants can be found in the larger land sized A2 farms (Moyo S. , et al., 2009; Moyo S. , 2011a).

Although the peasantry can be separated into six different strands of households, whose goals and interest in the cooperative varied, Chayanov (1991) still argued for its homogeneity based on the fact that the other groups of households are tough to alienate and hence are often found mixed and overlapped. The households include the *a*-type who mainly obtain income from trade turnover (returns on investment outside agricultural production), do not hire labour for their farming, have a more substantial influence in the countryside than all the other types. Then there, the *b*-type peasants that do not derive income from investments are productive and also hire labour, potentially exploited by the first group as wage labour or through unfair market interactions. The *c*-type have no non-farm investments, are productive (they can meet household reproductive needs from agricultural sources) and do not hire permanent labour (only casual labour at times) and are usually exposed to exploitation when they enter into commodity and labour markets. The *d*-type do not hire labour, are productive and do not sell labour, these use their labour on their farm exclusively, in addition to market exploitation, these are usually the ones targeted by capitalist exploitation through loans and credit above commercial rates. Then there is the *e*-type that do not hire labour, are productive but also sell their labour to other households, these are usually victims of all types of exploitation (in labour markets, usury/finance markets and outputs markets). Lastly, the *f*-type undertake subsistence production with their income based solely on selling their labour to other farmers. The last type usually will not join the cooperative because they lack the minimum resource requirements to participate in production (they cannot meet auto-consumption production). However, if initially provided with cooperative credit, they can transform from *f*-type into *e*-type and *d*-type which accentuates the importance of starting finance cooperatives ahead of all other types of cooperatives (summarized in *Table 2.1*). Thus, the majority of the peasantry found in the countryside at that time were concentrated in the *b, c, d & e*-type, and these were all ideal for participating in rural cooperatives (Chayanov, 1991, pp. 26–28).

It is this classification of the different farmer groups that prompted some scholars (Bernstein, 2009, p. 55) to argue that Chayanov did engage in debate with Lenin about peasant differentiation. Chayanov further noted that, besides profits from turn-over, income from other sources outside agriculture was held constant because a small proportion of peasants had this as an income source.

61

Table 2.1: Classification of the Chayanovian peasantry

Classification of the Chayanovian peasantry						
Type	Turnover income	Family labour use	Productive	Labour hiring		Sell of labour
				Permanent	Casual	
a	✓	X	X	X	X	X
b	X	✓	✓	✓	✓	X
c	X	✓	✓	X	✓	X
d	X	✓	✓	X	X	X
e	X	✓	✓	X	X	✓
f	X	✓	X	X	X	✓

Source: Adapted from Chayanov (1991, pp. 26–28)

To overcome the problem of the overlapping of categories highlighted earlier, Chayanov (1991, p. 31) further aggregated these six into two distinct types of households to which the 'social foundations of agricultural cooperatives rested upon'. Thus,

1) Those that exploit other farmers' wage labour for their income and are called capitalist market-oriented farms, whose interests or needs could be summed up using gross income, cost of labour (wages), gross profit, production costs and net profit margins.

2) Those that do not hire labour and or may rely on selling their own to other farmers, these are termed market-oriented peasants whose interests and needs are summed up using only the gross income and the cost of production.

Chayanov's differentiation reduced the peasantry into a bi-modal structure (having two forms, the poor peasant and the more affluent peasant even though he acknowledged the existence of intermediate types, but their numbers were not significant) (Chayanov, 1991, p. 31). The first class is essential, especially initially in the provision of capital resources which the second group can access as cooperative credit. Hence, we can then introduce the question of what level do farmers seek to produce? What is their motivation to produce? Theoretically, the peasantry is not too concerned about profit maximisation. Instead, Chayanov argued that the peasantry is concerned with utilising their labour up to a point where their marginal utility of output is equal to the marginal disutility of work

(S. Cook & Binford, 1986, p. 7). It would be a gross mistake to try and calculate profits for a peasant household in the same way as for a capitalist farmer because, unlike the family-farm, the capitalist farmer does not use his/her labour and his/her most significant goal is to make a profit (Thorner et al., 1986). Therefore, it is suggested that peasant household 'profit' should be termed 'payment for the labour of the peasant household.

Consequently, the ideal cooperative will depend on the structure of the peasants. In Zimbabwe, where significant proportions of the peasants are expected to be in the second group (market-oriented peasants), according to Chayanov, agriculture crop-specific cooperatives can strive. The cooperative mimics the household dynamics, and the cooperative can only grow in response to the growth in household incomes; thus, the progress of the cooperative is measured, not by the profits it makes, but by the increase in member income over time. The connection between the household and the cooperative was used by Chayanov to decide the sequence of activities and the types of cooperatives that should be formed as the cooperative develops. Just as in a household that requires (i) finance to (ii) purchase inputs, (iii) produce and then (iv) market or (v) process; ideally and indeed historically, cooperative activities should start with Credit cooperatives, then purchasing cooperatives, then marketing, then processing cooperatives and then eventually can take the form of multi-purpose cooperatives (Shanin, 2009, pp. 50–51).

The distinction between cooperatives & other social organisations

Cooperative organisations and other social organisations such as unions, clubs or churches have often been viewed as one type of organisation both at the institutional and organisational levels. Historically, they have even been hostile towards each other in terms of competing for members while in other cases, they have found a way to coexist. For example, in the agricultural sector, farmer unions viewed cooperatives as a way in which farmers could gain meaningful ownership and control over their products and processes of production (Cathy, 2017). Thus, cooperatives are substantially different from unions and other social not-for-profit organisations such as churches. The significant differences lie in how the organisations are financed, capitalised and to whom does the managing committee account. Also, in the manner in which the organisation interacts with the outside stakeholders (see also page 30). Not-for-profit organisations are accountable to the donors while cooperatives are accountable to the voting members. Although the internal structure of cooperatives and other not-for-profit organisations may be structured in the same way to improve democratic processes, another difference lies in the manner in which profits are redistributed to the community. For most not-for-profit organisations, they receive income from donations, and the organisation is structured not to make any profits. However, cooperatives get receivables from the sale of products and services, and profits are redistributed to the community depending on the number of activities done through the cooperative by each member.

Throughout the development of the agricultural social organisation, these different forms of cooperatives have existed in varying forms and have, to a more considerable extent, failed to transform the countryside. This is because their goals were not for the benefit of the members but the founders of the social groups in the case of unions, churches, for the benefit of the donors in the case of donor-funded social organisations and the state in the case of state-funded and controlled cooperatives. Collectives and early cooperatives in Africa were controlled by the state and mainly worked achieve the objectives of the state. Other donor-funded community organisations also worked within the confines of their funders, hence undermining the potential of the organisations. What I propose in this book are robust grassroots and genuinely farmer-controlled cooperatives that use the political apparatus to demand development from the state and fair market practices from the private sector.

Summary and conclusions

On the one hand, is the neoclassical and NIE theories that give the impression of cooperatives as a firm whose objectives is to increase member satisfaction by increasing their incomes and profits. These focused on making cooperatives more business-like while at the same time encouraging its role as a civil organisation that is responsible for making the government to formulate sound public policies. Therefore, neoclassical cooperatives emphasise on the free, individualistic market transactions of the cooperative entity while NIE highlight the importance of state regulatory institutions. On the other hand, Marxist theories on cooperatives viewed cooperatives as a means of controlling peasant capitalist production, fighting capitalism, and achieving socialism. While appealing in some instances, NCE and NIE theory of cooperatives has severe problems in the way it conceptualises the cooperative problem and hence in its solution of converting cooperatives into investor-owned firms. The same could be said for Marxist cooperatives which intended to replace capitalism with socialism which may not be feasible in the short run.

Chayanov's theory of cooperatives focused not only on theory but also on the practical sides. It is more suitable for understanding the new cooperatives understudy more than the other perspectives because farmers cannot avoid contact with the market (under Marxist) while at the same time they cannot afford to form private companies as individuals (under neo-liberal theories). Chayanovian cooperatives espouse solidarity-based values, which can be interpreted as an expanding community. Chayanov's theory of cooperatives is practical, dully targeted to agricultural sector unlike the generic NIE or the Marxist theories whose focus was on LMF and WMF (with less emphasis on peasant organisations, see Table 2.2).

Theoretically, federation-making may lead to the formation of a mega monopoly capitalist firm. There are advantages and disadvantages to having a strong federation; it needs to have an entry and exit point. For Zimbabwe, with over 70% of the population in the rural areas, a strong federation is desirable while for Japan with less than 5% as rural people, smaller coops make more sense. Another way to avoid degeneration can be to divide a grown cooperative to smaller cooperatives to revitalize the organization and make it closer to members. This is deliberately to go against the economy of scale. Federations can be suitable in the growth stage, while decentralization can be a good option in (or before) the degeneration stage.

Table 2.2: Summary of theoretical frameworks

Theory	Strength/Weakness
1. NCE & NIE	**Major Strength** • The drive to increase member income is important. • Helps to make farm production more business-oriented which helps increase economic production
1. NCE & NIE	**Major Weakness** • Wrongly assumes the problem of the cooperative is profit maximisation (which renders it a weak model for cooperatives) • No focus on social aspects of production • Low adaptability in Socialist national production models • The Cooperative lifecycle assumes the cooperative to be like any other private firm. • Concerned with increased production and pays less attention to the effects on social development
2. Marxist-Leninist	**Major Strength** • Recognises the need to farmer voice aggregation against the adverse effects of capitalist markets. • Recognises the need for economic democracy (majority control of local economies)
2. Marxist-Leninist	**Major Weakness** • No concise theory on the peasant cooperatives • Focuses on WMF and LMF • Inclined towards state-owned but member-controlled organisations. • Too abstract, low adaptability in modern-day Zimbabwe (compatible with Socialist production models) • Seeks to use cooperatives as a means to Socialism, less concern on the peasants themselves

	Major Strength
3. Chayanovian	• Practical • High adaptability in socialist and neoliberal national economic production systems • Designed for the agricultural sector. • Favours peasant owned and controlled cooperative organisations. • Concerned with solving more parts of AQ than the other two
	Major Weakness • His conceptualisation of the peasants has significantly changed. The theory needs adaptive modifications.

The next step in our story is to give an overview of Japanese agricultural and cooperative systems. This is important for this book to show the concept of state, community and market in the input and output supply system was put into practice. Particular focus will be on how agricultural cooperatives were structured, what linkages existed between the Japanese state and the community and to what extent did this nexus provide a better platform for the farmers to engage with the private capital markets. As such, the chapter that follows will discuss the historical background of Japan after WWII, then the evolution of the Japanese agrarian structure and the role played by the state in this process. A discussion on the structure of the inputs, financing and land markets is then provided.

It is also essential to understand rice productivity trends and output marketing trends because it is the staple food of Japan, as I shall do with maize for Zimbabwe in Chapter Four. We will discover the various types of cooperatives in Japan but focus more on the generic cooperative movement (supported by the state) and the relatively new cooperative movement whose influence ranges from local, regional to national spheres. This is extremely important to understand how the cooperative industry managed to succeed in increasing farmer incomes and community development (through amplifying farmer voices in government agriculture policy, for example) in Japan. We want to be able to draw lessons from the

Japanese experiences and see how these can be applied in the context of a developing country like Zimbabwe.

Chapter Three

State, Markets, and the Japanese Agricultural Cooperatives

> The ultimate goal of farming is not the growing of crops, but the cultivation and perfection of human beings.
>
> – Masanobu Fukuoka

Introduction

My grandfather died in 2013, three years before I moved to Japan to pursue my studies. He enjoyed talking a lot and had taught me how to listen. Every time he told us (the grandchildren) stories from his youth; I was always impressed. I wanted to be a good storyteller as he was, the way he structured his stories coupled with facial and hand gestures illuminated fireside stories. Only one of the aunties had the same skills in the clan, so I am not sure whether he genetically passed his 'superpowers' to me. Sometimes we would be listening to the news on his battery-powered radio; then he would ask my elder sister to turn the volume down because the news had reminded him of something, and he wished to narrate the story to us. In other times, he told us stories of how my father and his elder brother, had suffered untold injuries while turning young steers into oxen for draft power purposes. Or how my four aunties disliked it if meat was not included as a primary or side dish during dinner time. These types of stories were my favourite.

If he had been alive, it would have been my turn to sit him around the fire and narrate to him agricultural stories from the land of the rising sun. He was a good listener too, and stories of a faraway land would interest him surely. I imagine how I would explain the differences and similarities between Zimbabwean and the Japanese agricultural system, and how the former could learn from the later. I would have to describe the Japanese agricultural path, its first and second land reform, the current issues in Japanese agriculture and how these affected the development of cooperatives in Japan. His impression of faraway overseas lands was heavily influenced by the images of Europe, which is completely different from Asian landscape. Grandfather, just like many Zimbabweans, would find the various similarities between Zimbabwe and Japan uncanny.

Agricultural cooperatives and the Japanese land reform: Circumventing peasant radicalism

Japan has four main islands (and 4,000 other smaller islands) with Honshu being the biggest and enjoys high rainfall patterns throughout the year (Teruoka, 2008). Massive changes that occurred from as early as the 16th and 17th centuries are forever printed in the Japanese agricultural sector. Japan went through several socio-political periods that are significant for any scholar interested in understanding various aspects of contemporary Japanese life. It was once under the landlord Shogun ruling system. The most prolific and prolonged period was the Tokugawa Shogunate (feudal landlords) era popularly known as the Edo period (1600-1868) ruled by an influential family (see *Figure 3.1* on page 75). This period witnessed high rates of urbanisation, education, and trading, setting the ground for Japanese capitalism. It eventually became unsustainable and was replaced by the emperor/imperial system in the feudal period from 1868 to 1945 (Meiji restoration) before a parliamentary democracy was installed from 1945-present (Gluck, 1997).

There is a Japanese saying, 三人寄れば文殊の知恵 (san nin yoreba monju no chie) which loosely translate to 'Three people gathering can create wisdom'. The more people discuss various challenges affecting their lives, the more they can resolve them. Such beautiful proverbs bankroll the development of collective action in Japan. The development of Japanese cooperatives is divided into two periods, the pre-war and the post-war cooperative movements. It is critical to discuss these together with the Japanese land reform, and the resulting agrarian structure so as to appreciate the trajectory taken by the cooperative movement.

70

Pre-1945 cooperative movement

The Shogunate system disallowed unmonitored mobility; hence, the villagers were bound to their land. They paid land tax in kind and had hereditary usufruct rights to the land (Esham et al., 2012, p. 944; Ishida, 2002a). What cemented the groups together were the presence of the village head, whose job was to manage the activities of the villagers to ensure timely tax payments, thus when one or more of the peasants could not pay, it was up to the village head to rally the whole village to contribute and help pay the tax in full for the peasant. This encouraged teamwork and working together within the village and moved the cooperative culture henceforth. These villages were autonomous to a significant extent as the outside political structures did not meddle too much in its politics, so long as the tax was paid in full (Ishida, 2002c).

The peasants, as single household units, took an active part in this village set up primarily through the production of various market-oriented crops. Goods and services were exchanged using rice as the standard measure of exchange. The introduction of money towards the end of the Tokugawa period exposed farmers to higher rates of exploitation as they got deeper into debt with each credit facility they took. Eventually, as a way of addressing the constant attack on the peasants by the credit givers, the farmers formed cooperatives to gather their voices and be heard collectively (Dore, 2012). One of these earliest cooperatives were the *Shason* (rice stock association for emergency), *Mujinkoh* (mutual loan association), *Tanomoshikoh* (mutual financing association), *Hotoku-sha* (Association for the 'thanks for the favour') and the *Senzoukabu kumiai* (association for the administration of farmland) (Ishida, 2002c).

The success of these organisations in socio-economic development depended heavily on ethical leadership. The *Hotoku-sha* type resembled the Raiffeisen in its principles of economic benefit coupled with individual or community development. Some still exist in the villages of Japan today. Yet, this was way before the introduction of the Western type of cooperatives in the Meiji restoration of 1868. Credit associations, however, whose formation predates the 1900 cooperatives law, were formed through the unanimous decision of the feudal lords, and rice savings were

71

compulsory to prepare for a bad harvest. After the 1900 law, villages that adopted the German type of cooperatives had higher chances of forming credit associations (Esham et al., 2012, p. 944).

One of the most exciting things in the Japanese experience is the fact that they sent many scholars abroad to study other advanced societies and try to adapt what they learned to Japanese society. The first imported cooperative movement came in the form of Raiffeisen type (in the late 1890s) because the conditions in England were hugely different from that in the Japanese agrarian structure (Ishida, 2002c). At that time, both Germany and Japan had a considerable number of peasant and middle class (Birchall, 1997; Kappes, 2014, pp. 3–4; Kurimoto, 2004, p. 116). After deciding to adopt the Raiffeisen type, two scholars, Shinagawa and Hirata, proposed a law to establish credit associations modelled after the Schultze-Delitzsch in 1891 but it was considered too radical by the lawmakers (Ishida, 2002c). However, their efforts led to the transformation and improvement of the traditional cooperatives into progressive cooperatives with proper institutional structures.

The balance between government effort (as the supply conditions of cooperatives) and the farmers (as the demand conditions of cooperatives) was critical to cooperative development in Japan. The landowners and the owner-farmers played an essential role as they had both an economic incentive – to make sure that their tenants were producing enough for them to pay their rents, and for them to be more productive, and a social incentive – some held positions in the government and hence wanted to develop their villages (Ishida, 2003). Thus, leaders and upper-class farmers were critical in the formation of cooperatives; they were the drivers, often forcing the smaller farmers to pay share/owned capital. Eventually, the Ministry of Agriculture and Commerce passed the Act for the establishment of credit associations in 1900. This led to the creation of the purchasing, marketing, production (later called utilization) and the credit associations. It demonstrated that the Japanese cooperatives followed the German movement, where credit associations were disallowed from carrying other businesses other than they were registered (Asuwa, 1962). However, they were allowed later in 1906, after Germany had revised its cooperatives law in 1889 as well.

The 1906 changes to the Cooperatives law were followed by the 1909 revisions that gave provisions for formation of federations at both local and national levels to oversee the overall business of the cooperatives (Asuwa, 1962, p. 41). The third revision then followed in 1917, which allowed cooperatives to expand to provincial levels depending on their capacity. This marked a move away from pure cooperative principles because it limited the options of cooperatives one could join. In this same year, the Agricultural Warehouse Law was enacted that sought for the establishment of agricultural storage facilities in the rural areas. It was to be supported by a 20% government subsidy (the government would later avail 50% of the share capital in the formation of the cooperatives bank) (Asuwa, 1962, p. 43; Ishida, 2002c).

In 1921, there was another major revision in the Agricultural Cooperative Act which saw the establishment of national federations like the National Marketing/Purchasing Federation of Cooperatives or the Industrial Cooperative Central Bank Law. These revisions were a deliberate attempt to centralize the cooperatives movement, and eventually, amendments to the law were made to allow the appointment of cooperative directors. This meant the state now had substantial influence in the running and direction of the cooperatives. There would eventually be more minor revisions of the Agricultural Cooperative Act from 1921 to 1937. The control of the cooperatives bank through the directors meant the cooperatives had to follow the wishes of the state, and in doing so, they became a tool for siphoning out money from the peasants to the urban areas to fund industrial development (Ishida, 2002c). However, Ishida (2003) maintains that the development of these cooperatives benefited the farmers by bringing on to their doorstep, credit/savings and marketing services otherwise unavailable to all small-scale farmers.

Before the 1930s agricultural recession, voluntary landowners and wealthy farmers formed industrial cooperatives/organisations that focused on the provision of credit and other financial services. After the recession, the government installed constitutional provisions for the establishment of a comprehensive nationwide multi-purpose cooperative system that would extend beyond finance and credit. These could now do crop input and output marketing. The government's five-year plan (1933-1937) for multi-purpose cooperative development sought to make sure that cooperatives had covered all areas of Japan through converting almost all the villages and districts into cooperative units which would govern themselves (Yamashita, 2009). They were hugely successful and that the only opposition they faced came from the private companies that were losing out to the cooperative organisations now protected by law (Ishida, 2003). However, during the war, in light of food shortages, the government brought the cooperatives together with the political groups to form agricultural associations. The wartime Agricultural Organisation Law (1943) was created to amalgamate some of the urban and rural cooperatives to work together in the bid to be victorious in WWII. In this case, the orientation of the cooperatives moved away from serving the farmers to national interests — this era set back the development of the cooperative movement. The resulting amalgamations were better known as the *Nogyokai* organisation and were effectively not hard-line agricultural cooperatives.

Pre-1945 agrarian structure

The current agrarian structure is a result of two major reforms, the first occurring at the beginning of the Meiji restoration and the last one immediately after WWII in the mid-Showa period. Ogura (1966, p. 157) argues that in as much as these land reforms shifted political relations, they had little significance on the overall productivity of the farmers. Instead, it was increased use of technological and Green Revolutionary type of inputs that resulted in the rapid expansion of agriculture (given that there was no increase in the average land size held by each peasant farmer). However, typical land reforms affect not just political structures, but it influences peasant's socio-economy (cooperatives and organisations), the private sector, as well as the government itself. Therefore, a discussion of the development of the agrarian structure within the context of these two significant reforms is apparent.

The agrarian structure in the Shogunate era consisted of the feudal lords, tenants and owner-farmers. Feudal lords held approximately a third of the total agricultural land in the Shogunate era, which rose to 45% by the start of WWII (Ogura, 1966, pp. 152–153; Teruoka, 2008). These feudal lords also had control of the social lives of the farmers as they often discouraged luxury and made sure to reduce the prospects of a peasant revolt, for example by removing all swords from the community (only a samurai could own swords, and these bore allegiance to the feuds) (Dore, 2012). In the 16th and 17th century, there was a rapid growth in population and land under agriculture, and hence the land tax collected increased as well. However, in the 18th century, expansion of cultivated land stopped, and with it, the increase for the feudal lords' tax. This prompted the lords to try other ways of exploiting the peasants to maintain a steady tax base. The Shogunate then encouraged the merchants and moneylenders to enter into the land markets, i.e., own, lend, and speculate on land values, which further accentuated the plight of the peasantry. It is then that revolts and farmer agency escalated, e.g. *with their feet* by renting land from other favourable landlords or stopping farming altogether. Peasant unions and tenant unions were not yet recorded.

Figure 3.1: Socio-political periodization of Japan (1600-2020)

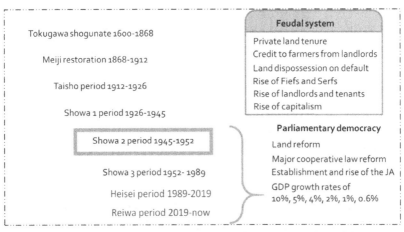

Source: Created by the author based on various sources (2020)

The arrival of the Americans in 1853 later resulted in the opening of Japan's borders to foreign trade. Japanese gold was grossly undervalued, which saw bulky exports to the USA. After realising that the price of gold was too low, the feudal lords reduced the amount of gold in their gold coins, creating inflation which mostly affected the peasants. The Shogunate lost its power over politics and economy, forcing the Meiji/emperor restoration (*Figure 3.1*). With this new system came in new land reforms, i.e., land was supposed to be surveyed and registered to individuals and they would be awarded title deeds called *chiken* (land rights). Therefore, land surveying was delegated to local villages, and the responsibility for tax collection was removed from the village level to the individual level. This was the birth of private property rights, which further accentuated the development of capitalism (Teruoka, 2008). From then on, we see the new state selling many public businesses to the private sector at subsidized prices, i.e., from industries, mining, construction. The establishment of the Bank of Japan in 1885 was the final condition necessary for the industrial revolution.

The abolition of the Tokugawa Shogunate feudal system in 1868 saw the establishment of individual farmer private property rights in the peasantry. This was the first land reform of Japan which eventually created the land-owning farmers, landlords (landowners who did not do farming) and tenant (farmers who owned no land) system. Three main categories of landlords developed; the absentee landlords, the village landlords (lived near the village), and paternalistic landlords (good landlords whose contract was guided by societal norms that proscribed the landlord from abusing the tenants) (Dore, 2012, pp. 23–33). These sometimes helped their tenants with issues that improved their general well-being and taught best practices in agriculture. It is these types of landlords that helped formulate producer interest-oriented cooperatives and farmers' associations/unions. Nevertheless, land rent was pegged over 50% of total production.

The role of the landlords has been debated in Japanese literature. Some argue that in as much as they had a positive impact on rural areas (formation of Cooperatives), it could have been better if the surplus-value of agriculture was not concentrated in a few. Although

76

the concentration of capital enabled expensive things like farm and irrigation refurbishments to be more accessible, this could still have been mobilized more efficiently without the presence of landlords. Japanese industrial capitalism predominantly depended on the rural population to replenish its labour force which often deteriorated rapidly because of the poor working conditions. Most of the people employed in fabric/cotton spinning factories were (mainly young) women who were 'sold' by their parents to these factories (Teruoka, 2008). The state maintained a huge fiscal budget that was mainly supported by the land taxes at first and subsequently by other various commodity taxes (soy sauce, alcohol, and sugar). Hence capitalism and imperialism in Japan were by and large financed by taxing the peasantry. In this respect, Dore (2012, pp. 47–50) asserts that, based on data from the Meiji period, farmer-owned productivity was overwhelmingly higher than tenant productivity, hence, supporting the argument that the landlords were able to use surplus-value to kick-start the industry.

By the time of the Showa period, it was clear that this agrarian structure could not sustain itself, the fall of the tenancy system was inevitable, irrespective of the war or the US occupation. Overall, land reform was pushed from four fronts. There were radical reform *Marxists & tenants*, then *landlords* – concerned with preserving their advantage through the political structures. The third groups were *bureaucrats & the army* supported by the *academia* and the press – mainly concerned with social stability. The last group espoused the *Nouhon-Shugi* (農本主義) – a pre-war ideology that believed that real development should be based on agriculture, they were nationalist in nature (Dore, 2012, pp. 54–63; Iwasaki, 1997).

Of interest is the emergence of farmer associations and groupings in the early Meiji period as they sought to unify their voices to address the ever-increasing land disputes, tenants, landlords and conciliation unions issues. The academia/intellectuals played a significant role in the formation of these unions, e.g., the Japan Peasant Union in 1922 in Osaka. The peasant unions were so aggressive that they competed for influential positions in the Diet, on the one hand, encouraged the formation of more collective organisations (including consumer organisation) on the other. The formation of collective organisations and unions also led to increased disputes and the need for secure tenure by the tenants (second land reform). Although riots/violent confrontations did not directly lead to the land reform itself (before 1945), it achieved a lot as in 1923 a law was enacted to allow for the reduction of rents in times of bad harvests and the recognition of fixed tenants was established.

The process of land reform (1945-1948)

Setting the land reform framework

The tenant & land questions have always been topical within Japan, from the Edo period through the restoration period, and even after parliamentary democracy was established. Market-based land reform was attempted in the 1926-1937 period in which tenants who wished to buy land (given that such land was available) could get a subsidised loan from agricultural cooperatives to purchase land (Kawagoe, 1999, p. 22). This initiative was, however, a top-down approach which was pushed by a reformist portion of the then Ministry of Agriculture and Commerce. The establishment of the Land Tenancy Conciliation Law to see to any tenant-tenant or tenant-landlord disputes was the first step in this reform. More importantly, the government established the Owner Farmers Establishment Law (1926) to oversee the speedy transfer of land from landlords to tenants once they had agreed on a price. The law provided for the cooperatives to give loans to tenants at 4.8% interest and that the government would provide a subsidy of the purchase price of up to 1.3% (Kawagoe, 1999, p. 23). Although it was successful in redistributing land, just like Zimbabwe's land reform in the 1980s (see Chapter Four, page 114), it was sluggish, and tenants purchased low quality and highly disaggregated portions of land.

With the defeat of Japan in WWII, the tenant question resurfaced. The release of political prisoners after the war meant that there would be a revival of the left-wing movement (which was evidenced by the sharp rise in the number of peasant unions) (Banno, 1997; Grad, 1948). Although much praise is given to the General Headquarters of the Allied Powers (GHQ). One way or the other, the land reform, or some form of it, was inevitable. The GHQ was

78

nothing but a midwife delivering a baby that was conceived much earlier on (Dore, 2012) as evidenced by the Land Reform bill of 13 October 1945 (Kawagoe, 1999, pp. 8, 27). There seemed to be confusion about the push for land reform within the GHQ as official documents and plans for post-war Japan drafted by the GHQ did not include implicit instructions about carrying out land reform, except the 'need to democratize' the rural areas to enable real development (Dore, 2012, pp. 130–131; Kawagoe, 1999, p. 22). There were possibilities that the GHQ was afraid that the land reform could open the way for communism.

In this phase, several laws were enacted to pave the way for land reform; for example, the conversion of rent payments to cash while simultaneously reducing the amount payable (Dore, 2012). Most importantly, a revision to the Land Reform Law was made to specify how absentee landlords and land leasing would be handled. The revisions had to accommodate the concerns of both the tenants and the landlords. The final effect of all the laws was to allow for the acquisition of *i*) all land of absentee owners, *ii*) all tenanted land owned by resident landlords that exceeded a certain threshold – depending on region, *iii*) all land cultivated by the owner-farmer that exceed a certain threshold, *iv*) any uncultivated land suitable for cultivation, and *v*) other residential land buildings or grassland rented by a tenant.

As part of the process, just as I later discuss in the case of the Zimbabwe land reform, three-tier land committees were formed which comprised representation of the tenants and the landlords as well as owner farmers. The committees had to administer the land reform at the village level. These comprised of five tenants (owning less than 33% of their farm), three landlords (cultivating less than 33%) and two owner-farmers (the rest). The prefectural level had the same structure except for double the figures of primary level and was elected by town and village members. At the national level, the government appointed a central land committee which had eight tenant representatives, eight landlords, two peasant unions and five university professors who mainly scrutinised administrative orders issued under the various enacted acts.

Carrying out the 1945 reform

During the process of reform, many problems were encountered. The reform was an attempt to change the land tenure rights of over 6 million people, and approximately 2 million of them were in opposition (Kawagoe, 1999, p. 31). The timing of the land reform was not ideal (some scholars Chitsike, 2003, p. 10; Matondi, 2012, p. 13 believe this to be the problem of the Zimbabwean case as well), it was too soon after the war, and there still was confusion in the countryside (and urban areas). Weak communication systems worsened the situation, shortage of paper, understaffing of land reform committees, poor working conditions (offices had no heating systems) and inflation was ravaging throughout the economy. Also noted was the fact that the remuneration to the staff of the land committees was too low, and hence poor farmers could not juggle committee duties and farming, and hence were discouraged from taking part. Those farmers who just managed to take part were more susceptible to bribery (Dore, 2012). As such, Land Committees often did not know or understand the limits of their power and jurisdiction. This led to many of the processes being left to the discretion of the local land committees. This situation was quite similar in the Zimbabwe land reform committees, as reported in Moyo (2007, pp. 108–113).

As already highlighted, the land reform was not a smooth process, which can also be seen through the various conflicts of interest which arose through the reform course. Land committees were biased towards landlord interests, and on many occasions, they had to be reshuffled, reorganised and in some extreme cases had to be disbanded and new committees installed. The same scenario was obtained in the build-up to the Zimbabwe land reform, in which the LSCF were firmly protected by private property rights and hence delayed and derailed the land reform through numerous litigation cases at the courts of law (Utete, 2003, pp. 50–51). Additionally, just as it later happened in Zimbabwe (where land occupations were driven from below in the initial phases but became top-down when it was formalised in 2001), the land reform in Japan took a top-down approach after the war in which the government dominated the movement. One of the reasons for a robust top-bottom drive (state

& GHQ) was to subdue the socialist and communist movements in the countryside (Dore, 2012, pp. 153–155).

Within the land reform process, class struggles emerged mainly in the enclaves of landholding status. Owner-farmers formed a social class whose goals was divorced from that of the feudal landowners and the tenants. The land reform also influenced the rate of farmer's union formation in the countryside (tenant, owner farmer and landlord unions). Here we see the power of collective action and aggregating voices as a form of agency. Because of the proportionate increase in the number of land dispossession cases, there was a corresponding increase in the number of agricultural unions from 1945 to 1947 (24,000 in 1946, 1 million by 1947 and over 2 million unions by 1949) (Dore, 2012, pp. 168–169). Although farmer's unions mostly attracted owner-farmers who did not eagerly support the reform, it was predominantly left-wing and hence also worked with the equally leftist Japanese scholars in this era (Ishida, 2003). The push for repossession of tenanted land even led to the formation of several landlord unions that helped landlords to appeal (in courts) against any land that they wanted to repossess from the tenants. This eventually presented itself as a threat to the whole reform such that the GHQ had to issue orders to stop and disband some landlord unions (Dore, 2012, pp. 171–173). This, though negative, highlights the strength of collective action.

Another similarity between Zimbabwe and Japan is that both did not destroy the large-scale capitalist farms as can be seen by the retaining of some landlord and tenanted land. The land reform law had provisions for each landlord to retain approximately one hectare of rented out land. The Diet (parliament) explained this, however, by arguing that due to illness, changes in family sizes and death, provisions for families to lease out land should be made. This was also noticed in the Zimbabwean case when the government left some commercial farms virtually untouched because they formed what was termed economically 'strategic farms'. Additionally, land rental markets have been witnessed in the past decade. For the Japanese case, we see the enduring influence of the landlord while in Zimbabwe, it was the case of a residual influence of large-scale farmers (backed by capital). Japan's reform was officially over in

1950, by then the country was experiencing inflation, and several tenants paid off all the amount of the land they had received from the landlords within a year or two (Kawagoe, 1999, p. 32). This fact was yet another similarity to Zimbabwean post-reform hyperinflationary scenario. I can imagine describing this to grandfather, and I guess he would ask if the inflation was anything close to what he had experienced in Zimbabwe from 2005-2009.

The post-1945 agrarian structure

The Japanese agrarian (second) reform is often described as complete and an enormous success because it managed to meet the bulk of its objectives. Before the land reform, the agrarian structure was said to be disadvantageous to the tenants, since their land sizes were too small to be profitable, and they always defaulted on their loans. Their land tenure was highly insecure (Kawagoe, 1999, p. 28). By the end of 1952, the reform had managed to successfully address issues of; overcrowding on land, unconducive/exploitative tenancy conditions, high farm indebtedness or high-interest rates on loans given to the tenants (in the context of widespread inflation), reduce the focus on industry and most importantly, reduce the government control of the farmer organisations without due regard for farmer's interests (Ishida, 2002a).

Another point of convergence between the two countries under study is that Zimbabwe land reform also sought to reduce overcrowding and poverty levels in the CA, and tenure insecurity in the resettled areas after the reform. For Japan, the primary concern was to prevent the return of landlordism. This is relevant for Zimbabwe because, in addition to the imposition of state tenures (which restricts market transactions in the A1, A2 and CA), post-reform fear of land grabs and farm dispossessions presented a risk to the reversal of the land reform. As shown in Figure 3.2 the agrarian structure of Japan was dominated by owner-farmers after the reform in 1949-1952, who also dominated the membership in agricultural cooperatives. In 1950, as a measure to prevent the reversal of the land reform, the government gave a provisional Order number 307 (until the establishment of the Agricultural Law). The order was later reconfigured into the Agricultural Cooperative Law (discussed on

82

page 92). It however focused mainly on landowners with less regard to the other forms of tenancy that existed.

Figure 3.2: Pre and post land reform agrarian structure (1941 – 1955).

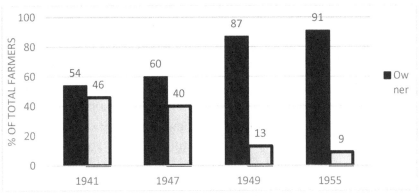

Source: Kayo (1977) in Kawagoe (1999, p. 54)

Although farmers now held land under private property, the state (through Order 307), the cooperative movement and the local governing body still controlled land market transactions. The market could decide the prices, but the law had restrictions on who had the right to purchase agricultural land. If landowners wanted to sell their land, the tenants had the right of the first refusal while corporations, foreigners, or non-farmers (farmer certificate holders) could not purchase farmland. Furthermore, not any farmer could buy land, and only those who were able to cultivate at least 0.3 hectares could buy land. There was a maximum limit on the land size that farmers could accumulate as well. Absentee ownership was strictly forbidden except when landowner stayed in the same village as his/her tenant. This land tenure system was maintained for a long time going into the 1990s in which the system became a problem for agricultural structural adjustments. The smooth operation of this law was made possible by the widespread cooperatives that were in place at that time.

The Post-1945 cooperative movement

As already highlighted, the government had a lot to say about agricultural development in Japan. This can be seen through the various laws and regulations that were enacted year in year out. After WWII, the government, under the instructions from the GHQ, dissolved the pre-war cooperatives by reassigning all the economic activities of the *Nogyokai* to the *Nokyo* cooperative. This was further made official by the Agricultural Cooperatives Law of 1947 and through a couple of other laws such as the Consumer Cooperatives Law (1948); Fishery Cooperatives Law (1948); the Small and Medium Enterprises Cooperatives Law (1949); the Forestry Cooperatives Law (1950); and the Credit Cooperatives Law (1951) (Godo, 2014). The reestablishment of cooperatives meant the reinstatement of the pyramid system in which there was; *i*) a vertically federated system based on the Food Control Law (see page 87); *ii*) unconditional consignment of primary cooperatives to their federations based on the Law of consolidation of agricultural, forestry and fishery cooperative federations; *iii*) a uniform business policy based on the detailed regulations for bankers (Ishida, 2002a). The *Nokyo* movement differed from the rest to the global cooperative movement because it was a government establishment and covered all farmers in the countryside (100% farmer membership). A perennial government-funded and controlled cooperative movement is undesirable based on the experience of the British-Indian cooperatives (both in Europe and throughout its colonies) in which the state influence never stopped but increased, the state sees and uses the cooperatives as rural developmental tool undermining individual capacity building (Schwettmann, 2000, pp. 4–5).

The reinstatement of the cooperative system after WWII brought in a new type of cooperative that could carry out multi-purpose functions. This meant one single cooperative could provide members with many different types of services and hence reduce the costs of participating in several different cooperatives. There are currently two types; multi-purpose and single-purpose cooperatives in Japan (numbering 2006 and 3363, respectively) (MAFF, 2015). The multi-purpose cooperatives had three things that made them unique; *i*) comprehensiveness- provided all services under one roof *ii*) full (compulsory) coverage- all farm households had full membership in the cooperatives; *iii*) territorial possessiveness – farmers had no freedom of choice of which cooperatives to join as the one that was in their area automatically became their cooperative (Ishida, 2002a), violating the principles of Raiffeisen cooperatives.

The comprehensiveness of cooperatives had its problems as it led to the cooperatives focusing on profitable ventures and pursuing them ahead of the needs of farmers. Also, the provision of several services reduces specialisation among the labour force of the cooperatives. Yamashita (2015b) argues that the cooperative in Japan was not as independent as is required under strict cooperative principles. He argues that the JA was formed from the organs of the wartime command economy for the top-down transmission of information and instructions from the JA *Zenchu* head-quartered in Tokyo (operating through the 47 prefectures) to the farmers. However, this does not explain why the group was an obstacle during the passing of the Trans-Pacific Partnership (TPP)[11] with the USA. This view ignores the fact that the poor farmers voluntarily supported the JA (and also the LDP party) after the war. It ignores the agency of farmers in the socio-economic structures of the country who often *vote with their feet*. If the government had so much power in the movement, how come it remains a challenge to pass agricultural reforms till this day? It appears the farmers still have an unbridled agency which stretches high into the halls of political power.

The government continues to try and reform the JA to clip its wings because it has become too powerful. In the case of Japan, where only 6% or less of the population are farmers, it is fascinating that these groups of people were stalling a policy as infamous as the trans-pacific partnership (TPP). What this means is that, in the case of Zimbabwe, where 68% of the population is rural (World Bank, 2019b; ZimStat, 2012), it should surely benefit the rural areas if they have the power to decide national policies. The biggest lesson from

---

[11] The Trans-Pacific Partnership (TPP) was meant to represent a 'regional free-trade' agreement among 11 countries in the pacific rim. This had been debated widely with neoliberal Japanese scholars (Yamashita, 2015b) arguing for it while socialist scholars (McMichael, 2014, p. 11) highlighting how it was an attempt to contained socialist China's growing economic influence.

the JA and its battle with the political technocrats and bureaucrats is that if farmers' voices are brought together, they cannot be taken for granted.

Chasing after rice: The development of Japanese agriculture

Japan, is a mountainous island, whose total surface area is approximately 38.7 million hectares (0.4 million hectares less than that of Zimbabwe), has only 16% of its surface area as arable land (Ishida, 2002a) (compared to Zimbabwe which has 40% of its land as arable land). This had a significant influence on the average land sizes of peasant farmers who had to practice intensive use of land, fertilisers, chemicals, and irrigation facilities on an average area of two acres of land (Nagatani, 2015; Ogura, 1966, p. 154). In the pre-war times, rice was the staple crop mostly grown on 59% of the arable land (and provided 57% of the energy requirements) (Kawagoe, 1999, p. 19; OECD, 2009, p. 13). Livestock rearing was minimal due to the absence of good pastures, and animals were stall-fed on grass from dykes, root crops and concentrates. One of the reasons why farming succeeded in Japan was because of the growing Japanese population, which depended on rice for food and hence kept the price of rice relatively high throughout. The abundance of labour also played a significant role in both maintaining a steady rural population while at the same time following an aggressive industrialisation path driven by fabric export (Teruoka, 2008).

The current population of Japan stands at 127 million people (World Bank, 2017). Japan's population growth has been declining in recent years because of a drop in the birth rate. A population structure with higher proportion of adults over the age of 65 years (retirement age in Japan) is a result. The Japanese society is slowly changing from mono-race, mono-culture and one traditional belief because a number of foreigners (also some 3rd and 4th generation Koreans, Vietnamese, and Philippines) have moved to Japan to find work. Some of these people provide labour in the agricultural sector. Japan has never been colonised (save for the USA 'occupation' 1945-1953) and hence has had a limited external influence (Ishida, 2002a), but it has imported almost everything from religion to art and culture.

86

It improved (through community participation) and fused it with local knowledge to make it suitable for Japanese society. It is led by a Prime Minister who is elected by the Diet (国会, Kokkai)[12]. Some various ministries and agencies help the Prime Minister to administer a legal framework and give subsidies to the 47 prefectures governments and 3246 local administrators classified into cities, towns and villagers (federal system) (Ishida, 2002a). The Ministry of Agriculture Forestry and Fisheries (MAFF) has control and influence in the regulation of agricultural cooperatives through various enacted laws.

For us to understand Japanese Agriculture, it is necessary to appreciate how the government engages with the cooperative movement through various policies and government-related programs such as the Food Control Act (1968) and the Agricultural Basic Law (1961).

The Food/Rice Control Act (1968)

A discussion of Japanese agricultural cooperatives would be incomplete without a brief discussion of the Food Control System (1968). This law and its institutional structures were formed to manage/ration food supplies during the WWII. The same structures were adopted after the war. It is vital to note that rice took prominence in the 'Food Control System' to such an extent that it would not be far from the truth to call the Act the 'rice control system' (Mishima, 1992, p. 43). Up until 1968, by law, rice supplies went through the government which had full control of all the market transactions. This was mostly through three institutional structures; i) the MAFF, ii) cooperative organisations comprising the producers themselves, and iii) the distribution network encompassing the wholesalers and retailers (also cooperatives) (Francks, 1998, p. 2). These three institutions (led by state) controlled the rice market. This policy was a departure from the pre-war arrangement of 'free markets'. The state took advantage of the already existing cooperative networks and farmer organisations to administer the program. Since

---

12 The Diet is the legislative arm of the government of Japan (highest organ of state power and sole law-making organ) composed of voted members of the House of Representatives and Upper House of Councillors (GoJ, 1947b).

the cooperatives were organised from the village level to the district, provincial and to the national level, it was easy and cost-effective to run the program.

Rice moved from the producers in the village to the respective village cooperative. It was sold to the cooperatives at a fixed price, and no other channels could be utilised. From the primary cooperatives, the rice was sold to the secondary cooperatives, then to the national level cooperatives and eventually to the government that later sold it to the wholesalers and retailers (Figure 3.3). At each level, the price was determined by the price determination board in consultation with representatives from rice retailers, government, national (JA-*Zenchu*), provincial (*Zennoh/Keizairen*) and village-level cooperatives (Francks, 1998, pp. 3–4; Mishima, 1992, p. 52). There were no quality or grading standards.

In this sense, the rice industry was vital in Japan as it employed vast amounts of labour and supported livelihoods. The rice system was sometimes labelled as bureaucratic and inefficient, but the system worked, and it kept prices higher and stable for a considerably long time (stabilised food security). The cooperatives at the local and national level became the most critical players in the industry, as such more income accrued to them. And since they were organisations owned and controlled by the farmers, the benefits ideally accrued to the people as well.

From 1968 until the late 1980s, the system benefited everyone and was used as an anti-inflationary mechanism. Eventually, the rise in incomes and demand, alongside the continuation of the rice controls encouraged the black market activities (*Yami-mai*) (Francks, 1998, p. 6). Better-quality rice was shared on the black market, while the rest found its way onto the government-regulated market. Rice grades had to be formerly developed. In 1969, it was regularized to sell rice to wholesalers from cooperative distributors, and this became to be known as the *Jishu* marketing channel (voluntarily marketed rice) (Figure 3.3). Although the *jishu* prices were based on the state prices, they were a bit higher, and as such, the state had to gradually increase the producer prices to reduce the price differentials in the two markets. This was pushed mainly by the farmers through the cooperative. An annual meeting was held between cooperative

88

and the Diet members to discuss issues to do with rice buying and selling. The cooperatives had established some small monopoly in the sector and were adversely affected by the opening of the *jishu* market. Thus, began the convergence towards liberal agricultural markets in a controlled or regulated way. The markets were not wholly free but had a combination of the state, the community and the business corporates who owned retail and wholesale outlets.

Figure 3.3: The rice distribution system during the era of the Food Control System era

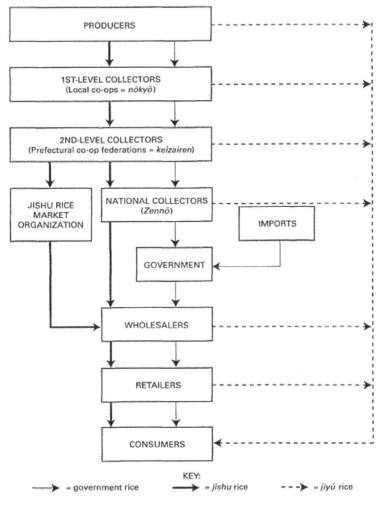

Source: Francks (1998, p. 4)

In the early 1990s, the pressure for the total removal of the controls mounted, the political leaders, bureaucrats, farmers (cooperatives) vehemently resisted. Although the law had stabilised the rice markets, there were outcries from outside Japan for the disbanding of the law since Japan had joined the GATT in 1955, and hence had to abide by the rules of maintaining unfettered market mechanisms (Kako et al., 1997; MAFF, 2018, p. 48). Francks supports this, writing that much of the pressure to open rice markets emanated from outside and the internal pressure was merely the final blow. Disputes with the USA over trade in beef and oranges before were well documented (Francks, 1998, pp. 1–11). The United States Rice Millers Association was one of the largest companies outside Japan to push for the free trade of rice between the US and Japan.

The politicians were weighing the cost of losing the market for manufactured goods with the benefit of protecting agriculture. The government chose the former and eventually resolved for the total but gradual opening of the markets. In 1993, before accepting the opening up of the agricultural markets, Japan experienced a cold season that negatively affected rice production (23% lower than average rice), this meant they had to import 2.5 million tonnes of rice from abroad and that the food control law had to be revised (Yamashita, 2015a). When the law was eventually repealed (December 1994), the farmers were the overbearing losers.

The new law removed the need to sell rice to the government, and with it, the state lost significant control of supply and demand, and therefore the ability to fix prices. Moreover, the control of the channels through which rice could be marketed ceased. The freely marketed rice was known as *Jiyu* rice. Although the cooperatives had lost out when the old law was disbanded, they continued to play an essential role in the new food law in which they became the major private channels through which rice was sold. The new law gave more freedom to the cooperatives as they could make their own decisions regarding the amounts and timing of offloading their rice on the market. This hurt the rice markets as they became highly unstable, thus increasing the chances of food insecurity. Cooperatives, who were now at the fore of these markets, faced higher market risks which were formerly shared with the government (Francks, 1998, p. 13).

Additionally, with unfettered market mechanisms came to large-scale commercial farmers who had the power to drive out 'less efficient' rice producers. The new law greatly disadvantaged those farmers who were represented by the cooperative, and an increasing number were driven out of rice production. This is the same situation which ESAP presented to Southern African farmers in the early and mid-1990s. Here lies another resemblance to Zimbabwe situation.

The Agricultural Basic Law (1961 & 1999)

The establishment of the Agricultural Basic Law in 1961 (and later the New Basic Law in 1999) aligned all other laws to do with agriculture together (George-Mulgan, 2005, p. 73; Kako et al., 1997). It put in place the institutional and procedural structures for administration of policy. While it made administration and implementation of government policy more manageable, it increased the state's stronghold on the agricultural system. The Agricultural Basic Law described and explained the exact type of activities that the government would support, and how farmers and stakeholders in agriculture should try to pursue such programs. This law also began to restrict the sale of land among farmers without the involvement of the state so as to avoid the entrance of private companies into the sector. Zimbabwe has been trying to forge a similar land tenure system since the 2000 FTLRP. The Basic law also had provisions to keep the gap between rural and urban incomes as small as possible. This was evidenced by an increase in the total proportion of the agricultural budget towards the management of rice and wheat (food control act) from 21% (1961) to 35% (1970). The structure of the agricultural budget followed the itemized list of the Agricultural Basic Law of 1961 (George-Mulgan, 2005, p. 184).

The Agricultural Basic Law of 1961 was updated to the New Basic Law in 1999. Although not wholly different from the old act, the main goal of the new basic law was the improvement of land tenure and farm organisation. The government wanted to improve productivity through amalgamating land into bigger, more commercially viable and manageable units (Article 2, clause 3). This they claimed was necessitated by the decline in rural populations and the need to reduce the disparity between farm and industrial wages/incomes.

Enlargement of land would be achieved through sales and leasing. Recently the agricultural reform law (2016/17) has taken this task head-on as seen through the works of the Organisation of Temporary Farmland Management (OTFM). Additionally, the latest law sought to maintain and preserve land for agriculture, thus, monitoring land-use patterns to adhere to environmental safety, and that land remains usable for future agricultural purposes. George-Mulgan criticizes this new law saying that it was another attempt to justify the government's intervention in the agricultural sector (George-Mulgan, 2005). Or put differently, an attempt to justify continued financial spending and subsidies on non-profitable agricultural ventures.

Other concerns of the new basic law include the need to ensure food self-sufficiency. The objectives were *i*) securing a stable supply of food *ii*) sustainable development of agriculture, and *iii*) promotion of rural areas. The new basic law is shaping contemporary Japanese agriculture as seen through the reorganisation of the MAFF into bureaus according to the precepts of the New Agricultural Basic Law. The state restructured and allocated even more money into agriculture to protect against agricultural trade liberalization. On the surface the government policy seems to be worried about the protection of the family-farmer land-ownership agrarian structure; however, George-Mulgan (2005, p. 132) argues that the Agricultural Basic Law has tried to consolidate land into more larger scale farms that are commercially viable.

Tools for peasant protection; Examining the Japanese Agricultural Cooperative Law

Agricultural cooperative laws from other countries across the world have witnessed several changes, especially after the 1995 ICA Manchester conference. As highlighted in Chapter Two, the rationale for updating and amending cooperatives laws is rooted in the desire to keep the cooperative system alive in light of widespread neo-liberal economic production models. For example, changes in cooperative law among developed countries were carried out in countries such as France (1992), Germany (1994), Japan (1996), Australia (1997), and Canada (1998) (ILO, 2001). Some changes enshrined new values and principles, others were more technical, for example, the sourcing of equity capital on the capital markets as a new form of raising capital. Although this situation is less noticeable in developing countries such as Zimbabwe, the level of capitalist production has forced lawmakers to pass laws that make it easy to convert some of the cooperatives into generally incorporated companies.

The Japanese Agricultural Cooperative Law provides a list of 13 types of activities in which cooperative organisations can be formed. These include; *i*) loaning funds for reproduction or better living conditions for members; *ii*) acceptance of savings from members; *iii*) supply of commodities for reproduction or better living conditions for members; *iv*) establishment of common facilities for reproductive or better-living conditions; *v*) establishment of facilities relating to cooperation in farming, sales, letting or exchange of farmland, or installation or administration of irrigation facilities; *vi*) transportation, processing, storage or sales of farm produce; *vii*) establishment of facilities relating to rural industry; *viii*) establishment of facilities for mutual insurance; *ix*) establishment of facilities for medical care; *x*) establishment of facilities relating to the welfare of the aged people; *xi*) establishment of the facilities for education to improve agricultural technology and farm management for members, or relating to the improvement of rural life and culture; *xii*) conclusion of collective bargains for the improvement of the economic position of members; *xiii*) and services that would accompany these 12 mandates.

Ishida (2002b) argues that the Japanese agricultural cooperatives law was and continues to be modelled around the German cooperative law as evidenced by the enactment of similar revisions as that of the 1973 and 1994 German cooperatives law revisions. The changes included introductions of a singular director and later in the 1996 amendments, the separation of the supervisory committee and the board of directors. These were being done in line with the changing economic environment (capitalism), size of cooperatives

93

and increased heterogeneity of members. The changes that were made to the cooperatives law itself were broad.

Further revisions and amendments to the Agricultural Cooperative Law were made in 2015. In revising the Agricultural Cooperative Act, the government recognized the role the cooperative had to play to help efforts of farm restructuring, which was being carried out by the OTFM (MAFF, 2018). The amendments re-enforced the need for the cooperative to reassert themselves in agricultural marketing of inputs and outputs, to make more profits and redistribute them according to the amount of business contribution by each member.

The State-Market interplay in Japan: Sites of negotiations

The role of Ministry of Agriculture, Forestry and Fisheries

The Japanese government (through MAFF) plays a significant and pro-active role in agriculture as well as in the cooperative movement. MAFF's current mandate is summarised as striving:

> to secure the stable supply of food; to develop agricultural, forestry and fisheries industries; to improve the welfare of the people engaged in farming, forestry and fisheries; to promote the development of agricultural, forestry and fishery villages and mountainous areas; to demonstrate the multifunctionality of agriculture, to increase forest production capacity, to protect and cultivate forestry, and to preserve and control fishery resources appropriately (MAFF Establishment Law, 2003 as cited in George-Mulgan, 2005, p. 53)

However, before the Establishment Law of 1999 (implemented in 2001), the institutional structure was slightly different from the current set-up.

MAFF was organised into bureaus equipped with a certain degree of independence and discretion from the central government. From 1945 to 1999, the MAFF had one secretariat, five bureaus (economic affairs, agricultural structure improvement, agricultural production, livestock industry production and the food & marketing). Then three agencies which were for food, forestry, and fisheries; additionally, there existed a food agency department which was organised outside the ministry but worked closely with it. The Economic Affairs Bureau had an agricultural cooperatives department (*Nogyo Kyodo Kumiaika*) responsible for the development and management of cooperatives. The same bureau oversaw agricultural trade policy in its international affairs department (*Kokusaibu*) (George-Mulgan, 2005, p. 47). This explains the disputes between the government and the cooperative on the food control system (see page 87) because both had power in terms of decision-making.

After the reforms of 2001, the structure consisted of five bureaus, and the number of agencies did not change (Figure 3.4). The Agricultural Production bureau was responsible for issues of production, distribution, and consumption of agricultural products (such as inputs, machinery and fertilizers) which ironically was also the responsibility of the Agricultural Cooperatives. This signified a point of convergence (or sometimes point of conflicts) between the ministry and the cooperative movement.

This bureau, together with the Management Improvement Bureau (responsible for land, labour, cooperative supervision, and management-structure of agricultural societies) formed the driving engine for the ministry of Agriculture. This meant that they were responsible for the supervision of the agricultural cooperative movement itself (see Article 4, Clauses 1 and 4–12 of the revised MAFF Establishment Law). Furthermore, to emphasize how important and influential the cooperative movement was for the ministry, there was a cooperative inspection department within the secretariat. As discussed in the preceding section, the government opted to promote the multi-purpose cooperative ahead of the single-purpose model.

George-Mulgan (2005, pp. 54–59) even suggest that the MAFF policy virtually 'took over' the lives of the people in the rural areas, because they were supervising and monitoring their social, economic and political spheres of agriculture and related industry. The government of Japan mediated in the agricultural sector through two main ways. Firstly, regulatory interventions which involved extensive implementation of legislative tools that dictated the flow of food within the industry done through such policies as the food control system (see page 87). Secondly, a direct market intervention which involved the direct involvement of the government in purchasing and selling (including imports) of rice, wheat, and barley through a monopoly of the market held by the Food Agency between 1942 and 1995).

Figure 3.4: The structure of the new MAFF internal structure
英文農林水産省組織・機構名

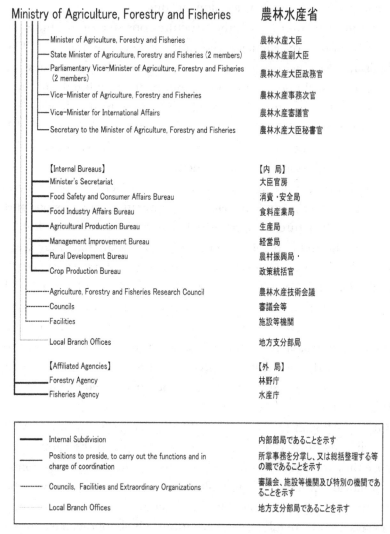

Source: MAFF (2020, p. 1)

The major disadvantage of government intervention is the fact that agricultural policy becomes soaked with subsidies which affect the efficiency of the sector. The Japanese define the cooperative as the private sector which gives different implications to agricultural market liberalization. The JA has been resisting the neoliberal and profit-oriented privatization pressure of deregulation for two to three decades now. Before 2001 (and before the 2015 self-reform), the JA colluded with the state through the strength of its structural organisation. The current role of MAFF and government is substantially changing from interventionism or protectionism to laissez-faire. The Japanese government is rich (developed country) and hence can afford to subsidize the farmers heavily; the same cannot be said of the Zimbabwe governments. The role of the Zimbabwe government also changed, but not in terms of 'from protectionism to laissez-faire', but it just ignored the movement altogether citing financial constraints.

Thus, cooperatives in Zimbabwe cannot survive if they base their sustainability on government subsidies and free finance. One of the most important differences or lessons we can learn from the Japanese experience.

Overview of agricultural financing in Japan

As briefly mentioned above, the Japanese government is rich and supports agricultural policy (subsidies, finance, research, and regulation). Government policy was funded through central and local government budgets. Subsidies form a considerable part of this budget (ranged from 41-63% in the period 1960-1985 and stood at 64.7% of the total allocation in 2002). Other sources of funds for the agricultural policy came from the Fiscal Investment and Loan Program (FILP) and funded various institutions in and outside the government (George-Mulgan, 2005, p. 86). In turn, the FILP was sourced from a combination of the private sector, other government businesses profits and the general taxpayers. In such a situation, as long as the economy was active and productive, there would always be enough money to fund such programs. This is also the most significant difference between developed countries and developing ones.

The financing of cooperatives, on the other hand, is done by the Agricultural Cooperative Bank System and the Norinchukin bank. The bulk of the funds for the cooperative activities (from agricultural to fisheries and forestry) comes from the JA cooperative banking system (Norinchukin, 2019). The Norinchukin Bank is supported, on the other hand, by member deposits in the form of savings.

Figure 3.5: Flow of agricultural funds within the JA cooperative banking system (2019)

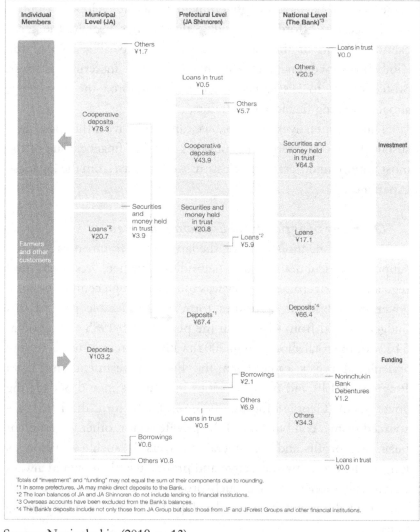

Source: Norinchukin (2019, p. 13)

The ACC is involved in two main activities *i)* to run the rural credit schemes and *ii)* they raise funding for the agricultural cooperative activities (Morozumi, 1993, p. 71). The ACC was responsible for the collection of savings in member households, lending those funds for agriculture and rural activities and transferring the remainder to the prefectural federations and making more money on behalf of the cooperative. A similar system should be developed in Zimbabwe. It does not make sense for commercial banks to be allowed to handle farmer's incomes when they are not prepared to give them loans. In Japan, the ACC is partly funded by the unit cooperatives. Some of the funds (more than a third – 40% in 1990) comes from investments made by the ACC through buying government bonds, stocks and private securities markets activities (shown as an investment in Figure 3.5).

The broad framework of agricultural cooperative financing can be summarized as consisting of income from *i)* the government financing institution (FILP, budget) and *ii)* from the Norinchukin Bank (collecting funds from their three/two-tier system). The agricultural finance system gets most if not all of its funding from the government and the cooperatives structures rather than straight from commercial banks.

Seven decades strong: The current condition of the national movement

The JA is the national cooperatives name for the multi-purpose farmer's organisation which is governed by the Central Union of Agricultural Cooperatives (CUAC or JA-*Zenchu*). Its formation was facilitated by the Agricultural Cooperatives Societies Law of 1947, which gave them the power to control and manage rural societies from the village level.

Ideology and institutional structure of the JA

There are two main types of decision-making systems utilised by the Japanese agricultural cooperatives: the unanimous and majority decision-making system. Unanimous or consensus decision-making is adopting a course of action based on the fact that all the people

99

involved agree to that decision. It is accepted through the need to foster group action. Opposing members are usually persuaded privately and promised a long-term benefit from accepting a particular decision (*Nemawashi*) (Ishida, 2003). In contrast, majority decision-making adopts a course of action based on what the majority of the people want. This style is consistent with the Western (German and Rochdalian) types of decision-making. Thus, an opposing member is given a platform to air his views and try to make people see things from his/her perspective. Informal discussions are discouraged.

The pre-war cooperatives based on the village structures leaned towards the unanimous rule as compared to the post-war cooperatives that focused on the majority rule system. These two approaches have implications on democracy because the former tries to encourage cooperation through seeking consent from the members that may oppose a given decision while the latter is more of a competitive approach that may divide solidarity since it creates winner/loser dichotomy. The group action which developed from the Tokugawa era; nevertheless, did so within the confines of equality and competition. Generally, anyone who was a threat to communal harmony and equality was ostracized by the village (Dore, 2012). The autonomy of the village then brought in a particular type of democracy which differed from the Western type. The Western-style (which is found in Zimbabwe) has more freedom but less equality, while the Japanese style had more equality and less freedom. And these two types require different forms and modes of management.

The structure of JA heavily relies on capital reserves rather than share capital. Although in the Japanese case, the proportion of share capital is stable enough from year to year due to the low mobility of members, it remains low to be relied on. Thus, capital structure, controlled by the directors and full members defines the success or failure of cooperatives. Although the institutional structure and ideology of the JA were imported from Europe (mainly German and England), it evolved considerably. Instead of following the British-Indian pattern of cooperatives where the people are 'helped' to create cooperatives with constant state financial and legal support, the JA movement was instead actively formed by the peasants themselves before the war, and then integrated as a quasi-arm of the government after the war. However, this did not mean the farmers were passive in the process. Because they were organised at a grassroots level, they made a deal for protection and assistance in exchange for economic and political support. As noted earlier (page 70), disbandment of agricultural associations after WWII was meant to reduce the level of communist ideology and the power of such fundamentalism within the movement for easy control of the movement by the state. George-Mulgan (2005) highlights how the aggressive state intervention through the MAFF compounded this problem.

The JA, under the encouragement of the government, adopted the multipurpose organisational arrangement to provide a one-stop-shop for inputs, outputs, machinery, insurance, and credit. This was ideal given the fact that farmers were small-scale and that they were settled on small land sizes; hence it would be costly for farmers to be in two or more different cooperatives that provide two different services (Ishida, 2002a, p. 20). Again, although borrowed from Europe, the Japanese cooperative movement took a different path in this respect and was successful to a greater extent. While single-purpose cooperatives still exist, often their members also belonged to the multi-purpose cooperative. The basic structure of the JA had three levels; the primary level, which was composed of cooperatives from the villages, towns and cities; a prefectural level which was composed of federated cooperatives that focused on the same type of activities; and the national level, which was composed of all the federated prefectural level cooperatives that focus on similar activities (*Figure 3.6*).

Other business sectors that fell under the control of the JA includes at the prefectural and national level, the Central Cooperative Bank (CCB-*Norinchukin*), established to handle banking and credit issues. Then there is the National Mutual Insurance Federation of Agricultural Cooperatives (NMIFAC-*Zenkyoren*), for insurance services; and the National Federation of Agricultural Cooperative Associations (NFACA-JA-*Zennoh*) which handles purchasing and marketing services.

Figure 3.6: The three tier-structure of Japanese Agricultural Cooperatives (JA)

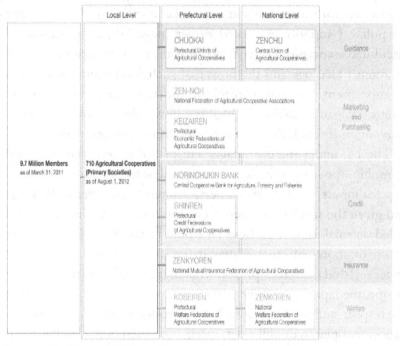

Source: JA-Group (2012, p. 3)

At the national level, the National Welfare Federation of Agricultural Cooperatives (NWFAC-*Zenkoren*) deals with medical care services and the Central Union of Agricultural Cooperatives (CUAC-JA-*Zenchu*) for general education, audit and consulting services (Ishida, 2002a, p. 23). All these organisations go under the JA banner and form part of a system whose goals and the national agricultural policy hugely influences functions, hence this has been the reason why it has often been used as a state institution in delivering rural development (Godo, 2014). From the establishment of the three-tier system until the early 1980s, there had been several criticisms of the system, arguing that it was very inefficient (Asuwa, 1962, p. 93; JA-Group, 2012; Prakash, 2003, p. 14). With the intensification of liberalisation and the gradual opening of Japan for international trade, JA and the government has consistently pushed to restructure the movement to a two-tier system (Esham et al., 2012,

102

pp. 944–952; Ishida, 2002a, p. 24). The resulting two-tier system will be devoid of the prefectural level because some of the functions would move to the primary level while some upgraded the national level.

The relationship between the JA and the Farmers

The JA-Zenchu uses the profits from banking (majority revenue of the banks comes from member savings), insurance and inputs/machinery sales to fund other cooperative programs such as extension and logistic shipment of farm products. The problem lies in the fact that the number of associate members has recently (from 2012) overwhelmed that of regular full members. Given that associate members are contributing more revenue, it is argued that the cooperative has shifted its core focus from the farmers to the non-farmers. This is the reason why the government suggested reforms that seek to reduce the benefits that associate or part-time farmer members can get relative to regular members (see page 104). While such institutions as the private sector and the government have used this as a way to apply pressure on the movement to reform (Esham et al., 2012; Yamashita, 2015b), other scholars argue that the cooperative needs to continue with credit and insurance activities to fund the cooperative non-profit making activities.

The primary cooperative management issues raised had to do with whether power should be vested in the management committee or the board of directors for more prominent cooperatives. Some countries advocate for power to be with the board of directors while others feel that the power should be with the management committee. The Japanese cooperative system has more power vested in the directors (who are usually attached to a previous government post). However, the ICA is more inclined towards the management committee (more independent and autonomous). Achieving independence and autonomy from the government will always be difficult since there is usually a low capacity to hire professional managers and poor working conditions for professional managers in rural areas (Ishida, 2003).

Active participation and involvement of members in cooperative affairs were identified as one of the critical things that cooperative movement should safeguard. The members should dictate the movement's affairs. Moreover, in improving member participation, people must understand the issues and the development path that they need to take. In doing this, education, training, and capacity building of the members becomes extremely important. Studies on the relationship between education and participation point out that educated and enlightened members could quickly and amicably find solutions to problems. Furthermore, the JA has been encouraged to reform itself in terms of increasing the income benefit from agricultural activities for its members while at the same time increasing the focus on core farmers as opposed to associate members.

The JA, reform, and the political economy of contemporary cooperatives

The MAFF is influenced by the Diet, whose members are connected to the agricultural cooperative boards. This has created an iron triangle which weakens the conformation of Japanese agricultural cooperatives to the ICA principles (see Ishida, 2002a). Over the past half-century, the Japanese agricultural sector has undergone various changes. The importance of agriculture to the Japanese economy has been decreasing while the imports of agricultural products, especially rice and soybean, has consequently increased. The complex agrarian structure and the productive industrial sector which drew huge populations from rural areas worsened the situation. There has also been an increase in the number of part-time farmers who now account for almost half of the 3.4 million farming households (MAFF, 2018). Full-time farmers produce more than part-time farmers (Esham et al., 2012, p. 952; Yamashita, 2009). Part-time farmers only grow rice while the full-time farmers have diversified their crops to include horticultural crops. However, both these types of farmers all have influential connections with the cooperatives and of late, the part-time farmers, have had their concerns take centre stage in cooperative business (MAFF, 2018; Yamashita, 2015a).

Just under 40% of Japanese arable land is put under intensive rice production while wheat and some horticultural crops were also grown in smaller quantities as an alternative. Livestock production was still limited. Up until the 1970s, Japan was food self-sufficient. As the number of people relying on agricultural production (rice) declined, so did the rural population. Currently, agriculture accounts for less than 2% of the GDP (GoJ, 2018, pp. 29–30; MAFF, 2018). This means that Japan is no longer food self-sufficient (self-sufficiency stands at 36%) and given the volatility of international food markets on which it heavily relies, it is at risk of waking up one day as a food insecure nation. The current agricultural policy of 2013-14 was put together to tackle, *among other things*, declining rural population and the increasing abandonment of farmlands mainly by ageing farmers.

As indicated, the JA has solid links with the politicians, and this was achieved through the Central Union of Agricultural Cooperatives (CUAC), which had a powerful influence in the Diet and hence in the MAFF (George-Mulgan, 2005, pp. 48–50; Ishida, 2002a, p. 23). The CCB and NIMFAC are the most profitable business ventures of the cooperative movement, and it is from the proceeds of these businesses that the cooperative was able to fund the rest of its programs. The CCB was initially co-owned by the government until 1962 when full control was transferred to the cooperatives. The heads of the CCB are usually former MAFF officials while the deputy-heads are usually ex-Ministry of Finance officials; hence this shows the active link between the most powerful arm of the cooperative and the most influential division of the government.

The agenda of the 1990s neo-liberal push meant a retreat of the state from the market. Hence, cooperatives that were mostly government-backed suffered. The JA was not immune. Its case was worsened by rapid developments in urban and industrial areas, which sucked all its labour. Other problems included an increase in world food demand, changes in diets, lifestyle and distribution channels (due to policy reforms such as the Food Control Policy) (Ichiya, 2016). Agricultural policy reforms carried out then were meant to reduce the power of the cooperative movement in lobbying against free and open borders for rice (see page 87).

In as much as there has been talk of resistance to reform by the JA, it has not been static. It has always changed its organisational, business and management structures depending on socio-economic conditions. The amalgamations in the cooperatives at local levels and the acceptance of associate and non-farmer members to the cooperatives (which shifted the dynamics and composition of the cooperatives) exemplifies organizational restructuring. In the same sense, the heterogeneity of the organisation has drastically changed. The business focus on credit and insurance was necessitated by an increase in the amount of money saved/loaned out to non-agricultural activities. The management aspect of cooperatives has changed as well with the increase in the number of staff members resulting from the amalgamation of the cooperatives at local levels.

In addition to agricultural input/output marketing and extension, the JA is also involved in activities such as welfare programs (youth and elderly programs, funeral services), fisheries, forestry, medical care, retail, insurance, banking, and credit. This is the essence of multi-purpose cooperatives. By the end of 2017, the overall cooperative movement had a turnover of US$145 billion per year, with over 65 million members in over 35,600 locations. The cooperative movement is virtually part of Japanese people's lives as approximately 37% (21 million households, which is seven times the number of households in Zimbabwe) of all households in Japan use the products and services of the cooperative consumer movement. Additionally, 25% (38 million people; which is more than twice the population of Zimbabwe) of the people are insured through the cooperative insurance system while 25% of total deposits are held in cooperative banks (JCA, 2017). What this points to is the fact that most of the current JA's profits come mainly from non-farm production sources such as banking and insurance (Figure 3.7). This enormous amount of money has attracted private businesses and lately international capital as well, thus convoluting the political economy of the state-led pressure for JA reform.

Additionally, many corporations have been pushing for the government to deregister the JA-Zenchu (including the JA-Zennoh and other prefectural bodies) as a cooperative and administer it under the Companies Act instead. This means the JA is caught between a rock and a hard surface and must find a way that ensures a win-win situation for them and the farmers.

Figure 3.7: Sources of profits for the JA-Zenchu (1970-2012)

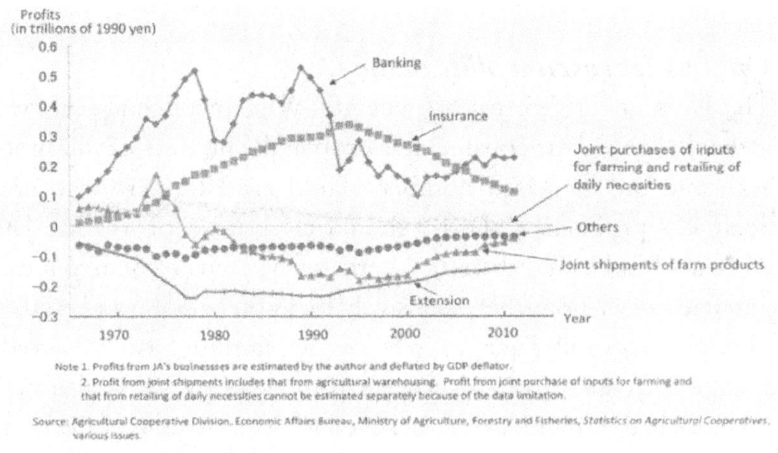

Profits
(in trillions of 1990 yen)

Banking

Insurance

Joint purchases of inputs for farming and retailing of daily necesities

Others

Joint shipments of farm products

Extension

Year

1970   1980   1990   2000   2010

Note 1. Profits from JA's businesses are estimated by the author and deflated by GDP deflator.
2. Profit from joint shipments includes that from agricultural warehousing. Profit from joint purchase of inputs for farming and that from retailing of daily necessities cannot be estimated separately because of the data limitation.

Source: Agricultural Cooperative Division, Economic Affairs Bureau, Ministry of Agriculture, Forestry and Fisheries, *Statistics on Agricultural Cooperatives*, various issues.

Source: Godo (2014, p. 5)

In 2015, JA conceded to the deregistering of only the JA-Zenchu from the Agricultural Cooperative Act, while JA-Zennoh was given the prerogative to decide on their own. For this compromise, the movement managed to defer the decision to limit the benefits that the all-important associate members could enjoy (Yamashita, 2015b).

The pressure to reform from some business bodies (scholars and such institutions as the Council for Regulatory Reform) who advocated for the removal of cooperative exemptions from the Antimonopoly Act (1947) is based on the rationale that *i)* the agricultural cooperatives no longer supports farmer's interest and that of the government, hence were behaving just like any other company that is concerned with making profits for itself *ii)* the cooperative does not need to be protected under the 'protecting enterprises with less marketing power' mantra because JA-Zennoh has a significant command of the whole market in Japan *iii)* the Act is being abused by some large-scale corporates to escape the Antimonopoly Act since companies can necessarily form or join cooperatives of their own (George-Mulgan & Honma, 2015, p. 113; Godo, 2015). On the other hand, pro-cooperative scholars argue that

107

they support a significant number of people and provide employment to a substantial segment of the population (GoJ, 1947a). Therefore, the removal of this exemption would significantly affect these people's livelihoods. It would destroy the prospects of keeping agrarian society, agriculture, and secondary industries afloat.

Options for sustainability of the JA

The JA is under threat to degenerate. While neoliberal scholars argue that in light of the considerable drop in the number of full-time farmers, and the increased number of old aged farmers who are relatively less productive, the JA should close most of its local JA departments (Yamashita, 2015b). Alternatively, convert them to local cooperative associations that exclusively focus on banking and credit. The cooperative will have to rely on its farmer base to avoid degeneration (Jossa, 2014). However, in the case of JA, whose ties with the original farmer base are believed to be weak, one of the best options that it has is to try and improve member incomes and benefits, while at the same time working to restore national solidarity. The government-backed reform supposedly encourages JA to serve farmers instead of being concerned too much about the business and profit-making. These cooperatives would be authentic and real cooperatives that base their existence on its farmer membership and internationally recognized cooperative principles. The cooperative movement needs to overhaul its entire business system, administration system and the institution itself. It needs to participate in value chains and improve its marketing system while being grounded in satisfying member's needs. This is not an easy task.

Summary and conclusions

In general, the Japanese cooperative movement has taken two broad forms. First, is the classical form which was found in the pre-war era, characterised by a centralized authority (often with heavy state influence) which aggregated consensus from its provincial and local members and then aggressively negotiated with capital. The second was the post-war form, whose '…outstanding feature is its enterprise centredness. This trend seeks to realize its aims exclusively through the corporate unit' (Banno, 1997, p. 118). The relationship between the Japanese government and the cooperative has yielded both positive and negative results. The most positive of these would be the resultant expansion and diversification of cooperative activities across all aspects of rural life. Thus, the government-controlled cooperatives and used them as a quasi-arm for rural development. At the same time, the ruling party needed the political support of the farmers and had to bring them to the negotiation table. On the other hand, through their advanced organisation at local and provincial levels, the farmers negotiated for the government to protect the agricultural cooperatives. This was a cosy relationship.

Study of Japan helped us understand how the JA overcame the free-rider problem. Membership is almost 100%, and people paid up part of their dividends into the cooperative reserve even after the law that made it mandatory to do so had been scraped off. This proved that the cooperative had the grassroots support of the poor farmers (diluting the top-down argument). The nature of the membership also extinguished horizon problems because cooperatives deposits are membership savings, and these form a considerable chunk of cooperative profits. Portfolio problems were also almost non-existent because Japanese cooperatives are multi-purpose and are involved in several income-generating activities.

Opinion leaders and upper-class farmers in cooperatives formation are relevant because they have an economic incentive (to make money) and social incentive (to see an area under their control progress), often they had to force the smaller farmers to pay share/owned capital. However, sometimes the power of opinion leaders within a cooperative may become overwhelming and detrimental to cooperative development; this is when the state can come in. The Japan Agricultural Cooperative movement is akin to the Raiffeisen type of cooperative (more equality, less freedom) as opposed to the Rochdale (which has more freedom and less equality). Each case requires different managerial modes which have implications for the study of cooperatives in Africa and Zimbabwe, where the British-Indian colonial cooperative systems had a massive influence. Thus, any recommendations for Zimbabwe must consider these factors.

In the next chapter, we review the path taken by Zimbabwe agriculture and the cooperative movement since the first cooperatives were instituted in 1909. I intend to elucidate, as done for the Japanese agricultural system, the role that the Zimbabwe state, the community, and the businesspeople played in the agricultural input and output marketing system up to today. The focus was on structure; the state-community linkages and the extent to which it fostered a better platform for the farmers to engage with the private sector. A description of this is necessary since it gives context that enables the mapping of the trajectory of the Zimbabwe cooperative movement. As such, the following chapter discusses the historical background to Zimbabwe during the colonial era, then the evolution of Zimbabwean agrarian structure and the role played by the state in the cooperative development process. A discussion on the structure of the inputs, financing and land markets is then provided as well as the sales market of maize, the staple food of Zimbabwe.

Chapter Four

State, Markets, and the Zimbabwe Agricultural Cooperatives

> You have little power over what is not yours.
>   — Zimbabwean proverb

Introduction

Grandfather was a typical peasant who had retired to the countryside after several years providing labour to the white-controlled industry in the city. He used to tell us about how he stood up for his rights against an oppressive 'Ian Smith government', as he used to call it. During his times, it was unheard of for a black man to stand up against their white man bosses in the industry, but grandfather had done it several times. When he could not stomach further injustices in the workplace, he decided to quit his job at Murray and Roberts Construction and become a farmer. Before he left, grandfather sued Murray and Roberts Construction for unfair treatment. You see, grandfather worked as a journeyman class one blacksmith for over a decade but was paid as a general hand. The lawsuit took over 12 months in a white man-controlled court; he lost the case and a lot of his entire savings in the same instance. When he decided to relocate to Hurungwe, he used proceeds from agricultural production which his wife was doing in the Zvimba rural areas.

After running away from the city, the same racial unjust policies followed him to the countryside. He again found himself producing within the dualistic agricultural system that disadvantaged the black farmers while subsidizing the LSCF. At Murray and Roberts Construction, grandfather hand-made several of the farm-implements we later used on his rural home including the scotch-cart, harrow, and wheelbarrows (they were stronger than any of the villager's). His scotch-cart had carried agricultural inputs and outputs to the markets throughout various agricultural policies from 1987

111

when I was born, to ESAP, the FTLRP and beyond. Sadly, it had carried some of the firewood we used for his funeral in 2013. Before age caught up to him, grandfather was well travelled to various parts of Zimbabwe, but he somehow had never told us of crossing the border to other countries. With the passage of time, travelling took a great deal of energy for him. After the FTLRP, it was perhaps my turn to tell him how the land reform had transformed rural and urban lives. If only he was alive, I would also tell him about my travels around Zimbabwe's resettled areas and beyond.

Given that maize is the staple crop of Zimbabwe (and was for Rhodesia as well), the general direction of the agricultural marketing policies was primarily based on the maize marketing policy. Just as in Chapter Three, where I followed the rice marketing system of Japan, I also follow the maize marketing system for the case of Zimbabwe. Thus, Chapter Four discusses the development of Zimbabwe agriculture from three perspectives: state agricultural marketing policy, land reform, and finally, the cooperative perspective. The chapter ends by discussing the contemporary relationship between the state and the cooperative movement in light of the new agrarian structure.

Agricultural cooperatives and the Zimbabwe land reform; Results of a radicalized peasant

The roots and nature of contemporary land issues in Zimbabwe are a direct result of several decades of land ownership activities dating more than one and half centuries back. The FTLRP should not be analysed as a static phenomenon, but as a continuum set in motion by socio-economic and political relations that were found before colonization; to the arrival of white settlers in the 1890s through to the 1960s (land alienation period); and finally to the market-led land reforms in the 1980-1990s (Moyo, 2011c; Moyo & Matondi, 2004; Muchetu, 2018, p. 71). The state controlled the land markets during the colonial period, and in most cases, the local indigenous populations were not allowed to purchase freehold land. The process of land alienation and land grabbing, which began with the arrival of the white settlers was reinforced through more laws in

112

the mid-1900s. It was only later in the 1930s that the Native Purchase Areas were created for a few black elites to participate in the land markets. From the 1930s, through a few laws (Land Apportionment Act of 1930, Land Husbandry Act of 1931, and the Tribal Trust Lands Act of 1965) oversize tracks of land were appropriated by the white-settlers and converted to private property lands creating a dualistic agrarian structure (Moyo, 1992). It is imperative to understand the structure and conduct of the land markets and the resulting agrarian structure because this has a significant bearing on the types of cooperatives that can be formed in these agricultural areas (see Chayanov 1991).

In Africa, cooperatives initially served colonial interest. African cooperatives were not bona-fide cooperatives because many lacked democratic member control; hence, they did not respect many of the cooperative values and principles. It is essential to examine the development of the cooperative movement in Zimbabwe and try to understand how it fared against various challenges along its growth trajectory, just as I did for the Japanese case in Chapter Three).

Pre-independence cooperative movement

Cooperation is part of society and can be said to have existed within the concepts of *ubuntu*[13] which formed the basis of the socio-economic organisation of the pre-colonial Zimbabwe (Samkange, 1980). However, 'formal cooperatives as they are known today began in the colonial era to support white-settler commercial farmers (LSCFs) and tackle output marketing and input supply issues for white farmers. The first cooperatives were formally registered in the aftermath of the Cooperative Agricultural Act of 1909 which formalized white-settler farmer's cooperatives.

In addition to the discriminatory dual markets in grain and livestock markets between whites and black agriculture in the 1950s, the settler government decided to intensify their control of the black agrarian sector by encouraging the formation of cooperatives

---

[13] The concept of *ubuntu* recognises the equality of a human disregard of the strength of kith and kinship. *Ubuntu* (*unhu* in Shona) is usually used in the axiom '*munhu munhu nekuda kwevanhu*', which means a person is only human if they recognize the humanity of others (Samkange, 1980).

113

(Masters, 1993; Muir-Leresche & Muchopa, 2006; Weiner, 1988). The Agricultural Cooperative Act of 1956 institutionalized this move under the Cooperative Societies Act (chapter 193) on 15th October in the African purchase lands of Mashonaland West Province, Chitombogwizi area. This came as a recommendation by the African Production and Trade commission which had been previously set up in 1944. The 1956 Act was based on the Ceylon Cooperative Ordinance of 1922, which was in turn modelled around the British-Indian Pattern of Cooperation of 1904 (remodelled in 1912 and 1946). The Act sought to i) organize bulk purchases of farm inputs ii) to provide marketing options for the surplus production that peasants were producing iii) to arrange transport for the members and reduce transaction cost. The Agricultural Services Cooperative oversaw these activities. Unfortunately, just like other economic sectors, cooperatives were also affected by the geopolitical and economic shocks that hit the globe, such as the great depression, the world wars and the liberalization of the markets. The sanctions on Rhodesia of 1963, the liberation war and the economic adjustment programs had its toll on the system.

Initially, the cooperative movement was so successful that provincial cooperatives had to be formed to provide centralized services to the primary cooperatives. Theoretically, the British-Indian model envisaged limited state control in the cooperative movement as it grew bigger and became self-governing. This model meant that the government would be involved in the guiding of cooperative activities as an advisor through extension, help craft by-laws, and teach cooperative principles to the developing societies. However, the experience in subsequent colonial states, legislative enactments substantially increased the power of the Registrar of Cooperatives. The government chose an officer who ran and managed all cooperatives businesses and performed most of the technical services and activities required by the cooperatives (World Bank, 1989, p. 5).

Post-independence cooperative movement

After attaining independence, the state officially followed socialist policies (a mix of state intervention and market-heterodox policies) and hence turned to cooperatives as a rural development

114

model. This was evidenced by *i*) the creation of cooperative-friendly legislature, *ii*) preparation of a cooperative policy paper in 1983 which detailed long term plans for cooperatives, *iii*) increasing the support and the staff to supervise cooperative development (though not sufficient), *iv*) establishment of the Ministry of Cooperatives (which survived for only a year) (World Bank, 1989). In 1980, given that the government had put hope on the cooperative movement to help in rural development, the cooperative was given several vital roles (such as input distribution, organize bulk purchases of outputs, and transportation). The Agricultural Cooperative Act of 1956 was administered in the Ministry of Cooperative, Community Development and Women's Affairs. The state also established a network of depots in the remote areas to make supply and purchase of small-scale produce easier (Bratton, 1987, p. 194).

Eventually, it proved too much to handle for the 1980s movement because of inadequate financial and institutional support from the state and private sector. A 40% cooperative mortality rate was recorded. The cooperative movement did not have enough trained staff, no storage warehouses and transportation facilities – it had to build these from scratch. There were approximately 343 registered agricultural service cooperatives for the 70,000 households at independence, and by 1987, this number had grown to 642 covering 125,000 households (World Bank, 1989, p. 5). It is not the establishment of cooperatives that determines success, but it is their management and the ability to produce agricultural commodities sustainably. Sustainability was only possible if both state and the cooperative movement had equitable control of agricultural pricing mechanism for agricultural inputs and outputs.

Additionally, other players who wanted to benefit from the agricultural markets influenced the success of these cooperatives, including the government itself. Thus, private companies that supplied the same market presented competition while the expansion of the GMB also had its negative effect. This showed government policy inconsistency since it was both promoting cooperatives and at the same time competing with agricultural cooperatives in purchasing grains.

Establishment of collectives

The collectives gained traction in the state policies because many African governments adopted socialist economic production models following the likes of China, Russia and Tanzania. However, the government did not abandon other conventional types of cooperatives. The prescribed role that the government was supposed to play has not changed even in the current cooperative societies act (Makochekamwa, 2015; World Bank, 1989, p. 4). At the peak of the government control, just as in China, collectives/communes were formed under the land resettlement program (under the name Model B). These were developed along the lines of the cooperative model, but the unit of production was not the individual within the cooperative, but it was the cooperative itself. Such is the nature of the collectives, and it is this nature that renders collectives unstable and unsustainable.

Table 4.1: Growth of agricultural service collective and cooperatives (1956-1987)

	1956	1960	1965	1970	1975	1980	1987
Agric service coops	2	21	169	283	310	343	642
Farming collectives	-	-	-	-	-	1	312
Other non-agric coops	-	3	7	18	30	32	848
Totals	2	24	176	301	340	376	1802

COOPERATIVES IN 1980

36% *Agricultural service cooperatives*

17% *Farming collectives*

47% *Other non-agricultural of cooperatives*

100% *Totals*

Source: (World Bank, 1989)

Farmers were supposed to form groups, register and apply for land as a collective. About 1100 collectives were created between 1984 and 1987 and consisted of approximately 35,000 members (mostly war veterans). However, the bulk of them were never registered and hence eventually did not get land. The number that managed to get registered amounted to 312 (with 13,000 members) to which only 93 managed to receive land (5,000 members) (World Bank, 1989, p. 6) (Table 4.1). Again, the WB argues that the reason why these failed also was that they were not built and supported (financially) to hold the overwhelming number of people or members that it eventually attracted. And also, because the state had too much power and control in their management.

A recent study discovered that collectives created during this time are still in existence and that some had increased in numbers (312 in 1989 to 400 worker and producer collectives in 2015). The collective membership had increased to approximately 18,000 from 13,000 in 1989, and most of the members were women (Makochekamwa, 2015, p. 21). The collectives are currently organised under the Organisation of Collective Cooperatives in Zimbabwe (OCCZIM). Although support for the cooperatives and the number of trained staff was limited, the cooperatives grew in the first decade of gaining independence. The 70,000 members accounted for 9.8% of the small-scale farmers in 1980, and this rose to 135,000 total cooperatives and collectives by 1987 (19.1% of the small-scale producers). This represented more than 100% increase in cooperatives in this decade.

The Processes of land reform (1980-2018)

Land reform attempts between 1980 to 1997

In the first decade after independence, the government sought to access and control land within the framework set by the 1980 Lancaster House Conference (LHC) agreement[14]. From 1980 to 1985, the sovereign state was still in its most infant stages, thus understanding the land tenure system it had inherited proved to be a herculean task (Herbst, 1988). In addition to the Communal Lands Act of 1981, the government set out the Land Acquisition Act (1985) to speed up the market-based land reform. This Act secured the right of first refusal to the government for any agricultural land to be sold. This had a limited effect on the rate of redistribution because land supply was deficient, of low quality, and priced artificially higher. The

---

[14] The LHC agreement was held in London (December 1979) to bring an end to the armed struggled between the Rhodesian settler army and the black guerilla freedom fighters. It was an agreement to a cease-fire, renounce use of force for political gain, peaceful post-cease-fire election campaign and a pledge to accept the outcome of the elections thereof (Davidow, 2019).

land redistribution in the markets (through state-mediated market mechanism) was profiting the white-settlers more while delaying land redistribution and re-establishing the white-settler control in both land markets and financial markets. The LHC agreement proved to be the most significant impediment to a faster land redistribution system between 1980 and 1990.

A few black elites managed to purchase land from the market through private loans secured from such institutions as the Agricultural Finance Cooperation (AFC). However, the rest of black Zimbabweans, the traditional authority and Zimbabwe National War Veterans Association (ZNWVA/WVA) were excluded from the land markets, and their agency to access land was not yet radicalized (Sadomba, 2011). It is noteworthy that women's access to land was low at that stage. Even in the white-settler farms, women-only-owned land stood at 5% of the total title deeds. The funding from the UK, Germany and the USA had not flowed to the government as informally agreed during the LHC (not entrenched in the agreement) (Masiiwa, 2005, pp. 217–218). By 1988, the flow of funds had drastically reduced, and only £44 million had been received amid increased conditions set by the Conservatives of the UK government. In addition to setting the price of land, the LHC agreement gave the white-settler the right to choose in which currency they preferred to be paid in. This further complicated and delayed land market transactions given the shortages of foreign currency that was rampant at that time. Although some scholars (Kinsey, 2004, p. 1671) underplay the effect of the LHC on the pace of land reform, it stunted the progress of land redistribution from 1980 to 1995. Only 60 thousand families were resettled against a target of 160 thousand households on 2.1 million hectares (7% of the arable lands) by 1990 (Masiiwa, 2005, p. 218).

The LHC agreement expired in 1990, and land reform was expected to speed up; however, the adoption of ESAP made sure that this did not happen. The effect of this in the land market was a U-turn in power relations from state-led redistribution to market-based/white landowner-controlled transactions which were worse than the LHC agreement. At this stage, the government was uncertain of the implications of continued land redistribution (the

118

Land Acquisition Act of 1992 which sought to acquire land compulsorily), while at the same time following a market-based economic mode of production. The Land Apportionment Act had much potential to accelerate the land reform had it been supported by the local farmers, donors and the international community. Instead, it was ridiculed. The IMF and World Bank recommended re-focusing of support to large-scale commercial production. Land redistribution was shelved during ESAP as focus shifted to the implementation of the economic reforms under the supervision of the Bretton Woods institutions (Moyo et al., 2014, p. 2).

Moyo and Skalness (1990) therefore stresses that: *i)* the purchase of land by influential party elites, *ii)* the unity accord of 1987, and *iii)* the continued economic crisis extinguished any thoughts of state-led radical land reform. By 1995, the prospects of a state-led reform looked grim. By 1996, we start to see agency coming from below, increases in peasant's rate of squatting in commercial farms, land occupations, widespread protests, armed confrontations and resource poaching as a form of agency against a state unwilling to do a redistributive land reform (Moyo, 2000, pp. 10–11; Moyo & Yeros, 2005a, pp. 182–186). By 1997, ESAP had caused disaster and untold mayhem in the land markets and to the lives of the rural poor by increasing differentiation in land use, labour, and commodity marketing. More black-business capitalist joined the white-settlers in their quest to access land. CSOs and CBOs were silent in this stage. The farmer organisations were also quoted as useless in collecting the peasants' voices as most were formed by white master-farmers who opposed land redistribution (Masiiwa, 2005). Isolated and low-intensity farm occupations/squatting that had started in the 1980s (by war veterans and non-war veterans, peasants, traditional leaders and farm workers) and were heavily repressed by the state then, had kept the land reform agenda alive (Sadomba, 2011). My conjecture is if there had been grassroots socio-economic organisations in the rural areas, amicable land reforms would have occurred or if not, the radical land reform would have happened earlier than it did.

119

Land markets and reform policies, 1997 to 2010

Although ESAP was officially abandoned in 2001, the government had started to move away from it from 1996 onwards. The budgetary support from the UK had officially expired. The new Labour Party made it clear that it was not going to support further land reform programs through the infamous Ms Clare Short letter (Secretary of State for International Development). The intensity of land squatting, inversions and occupations were picking up pace. The WVA which had been formed in 1989 took it upon themselves to initiate restructuring of land markets. Between 1996-1997, the government used the provisions from the Land Acquisition Act (1992) such as land under-utilisation, multiple farm ownership, derelict land, absentee farm-owners, and proximity to Communal Areas (CA) to identify and redistribute 1471 farms. Sadly, the program failed because it was implemented within a liberalized market which respected private property rights. Just as in the 1993 and 1995 court cases against the state's compulsory land acquisition program, the landowners challenged and won back 40% of these farms.

It was at this stage that farm occupations intensified beginning with the famous occupations in Svosve village in 1998 which spread to other areas such as Manicaland, Masvingo and even Matabeleland (Sadomba, 2011, pp. 98–105). This time around, the peasants – through the WVA – had amassed political connections and were not facing state repression anymore. While other scholars viewed this social movement as the works of powerful elites in the ZANU PF party, others argued that it was, in fact, the WVA who had established itself with influential elites (Moyo, 2000). The government, realising the extent of the land occupations and pressure for land by the poor, had co-opted it into its land acquisition agenda (as we saw in the case of Japanese government and the GHQ).

The 1998 donor conference was subsequently organised to try and mobilize funding to the Zimbabwe government so that it would pay compensation. The property rights (enshrined in the constitution in 1980) were still being protected under laws and delayed the compulsory land acquisition process much to the chagrin of the state. In the meantime, WVA and elite political leaders further fuelled

120

radicalism. The donor conference resolutions crumbled. By 1999, GoZ still could not smoothly redistribute the remaining 60% of the 1471 farms as more court challenges came. This unified the peasants, the state, WVA, party officials and traditional authority to push the land reform agenda forward. For the first time since independence, the WVA had amassed adequate power to challenge the inequitable land distribution (see Table 4.2). Once the peasants had a unified voice, we see the power of social organisations and the ability to link with the state apparatus in action for the first time since independence.

Table 4.2: Source, type, and effectiveness of land demand (1997-2020)

Period	Sources	Demand type	Influence
1997-2007 State controlled	White settlers	Tenure security	Low
	State	Redistribution	High
	Peasants	Redistribution	High
	ZNWVA	Redistribution	High
2008-2020 Re-liberalization	Foreign Capital	Tenure security, Leases, purchase	High
	State	Redistribution	High
	Peasants	Redistribution, tenure security	Medium
	ZNWVA	Redistribution, tenure security	Medium

Source: Adapted from Moyo (various writings); Muchetu (2018, p. 76)

In the wake of failed donor's conference, failed land reform and a rejected referendum (1997-2000), small cases of land occupations, farm inversions and squatting then developed into full-scale occupations that would last until 2003-2004. The peasants and WVA forced the state to implement the FTLRP through a new Land Acquisition Act of 2000, which finally removed the need to pay compensation for the acquired farms. Consequentially, a state-led economic model of production was adopted, which saw the GoZ reasserting authority throughout the agricultural markets, from input distribution to output marketing.

Although the WVA were already carrying out land occupations, officially, the FTLRP was launched in July 2000 as part of the second phase of the land reform (Sachikonye, 2005, p. 33). It was a completely different creature from the previous land delivery systems. It was ideologically different in that it no longer respected the property rights held by the white-settlers. This reform was institutionally supported by the state as seen through increased constitutional amendments to allow for no compensation, remove legal challenges and protection of the land occupiers from eviction (e.g., Rural Land Occupiers Act of 2000). Some of these acts had to rely on presidential decrees vested in the 'Presidential Powers Act' to be passed.

Moreover, the FTLRP was to be taken on an accelerated manner with speeding up of land identification, planning, demarcation, and resettlement of the people. However, just like the previous reform program, the FTLRP was hinged on achieving equitable land ownership, poverty reduction, increased productivity which formed the socio-economic objectives of the reform. Furthermore, it targeted the decongestion of CAs, improving their land access, and the formation of an indigenous commercial farming sector (just as in the Japanese land reform).

Most of the land seekers who took part in the land occupations were later officially given usufruct rights (permission) under the A1 resettlement tenure model with land size ranging from 1-30 ha depending on NR. The other resettlement tenure system, the A2 model was for those that proved that they had the capital means to utilise more land. There are four forms of landholding in this model with the small-scale, medium-scale, large-scale and peri-urban (subject to land size and NR). The A2 model beneficiaries hold leasehold tenure title. The FTLRP also left some former white-owned LSCF and some large corporate farms untouched because they were deemed as strategic farms; these still hold freehold tenure title.

Emerging land markets post-2010
The FTLRP did not wholly eradicate land ownership inequalities as evidenced by a persistent local and international demand for land post-2007 global food-crisis. The FTRLP nationalised some LSCF land, but some LSCFs were not acquired and remains under freehold title. These are subject to market transactions. However, new forms of local and international land deals have emerged post-reform signalling a new dimension in the land markets.

122

The emerging markets are characterized by large-scale land deals[15] as well as localized land sharing, leasing, and renting. The most recent example of this 'new scramble' for Zimbabwean land (Moyo et al., 2012) is the concentration of land for sugarcane production aimed at bio-fuels production where some farmers had to be moved to accommodate such projects in Mashonaland West, Masvingo and Manicaland. Large-scale land investments have also taken place, being disguised as contract farming or out-grower systems as evidenced in some cases in Chiredzi district (Little & Watts, 1994; Mazwi & Muchetu, 2015). Thus, pressure for the rationalisation of the shortcomings such as allocation of large farms, multiple ownership (peasants' side) and land compensation (dispossessed white farmers) will drive land demand/debates going into the future (see Table 4.1). The government has recently (July 2020) announced a US$3.5 billion payment to former settler farmers for the land that was dispossessed during the FTRLP sparking a whole new debate on the land question.

The emerging land markets pose a threat to the gains of the land reform and thus give scope for better organisation and cooperation as a bulwark within the rural spaces. The emerging agrarian structure and its challenges are ripe for the cooperative model.

The post-reform agrarian structure
The result of the FTLRP has been the focus of debate for the past two decades. Scholars and government policymakers grapple with the task of fully understanding the nature and socio-economic implications of the radical land reform. By 2010, around ten million hectares of land had been redistributed to over 170,000 households under A2 (commercially-oriented: 13% of the land) and A1 (small-scale: 79% of the land) settlement models (Moyo & Nyoni, 2013, p. 202). This was ten times the land and 2.5 times the number of beneficiaries as compared to 1980-1999 reform, all done in a quarter

---

[15] Large-scale land deals refer to land acquisition of larger tracks of land (usually greater than 200ha or two times the median landholding in that territory). Recent large-scale land deals of investment have marked a renewed scramble for African land which has often seen many peasants land being dispossessed through the state apparatus. The beneficiaries are international investors of tourism, wild-life preservation and bio-fuel production (Moyo et al., 2012).

123

of the time (Moyo, 2005a, 2011b; Muchetu, 2018). Exclusively using economic variables to measure the impact of the reform is inadequate, and there is a need for more profound social and class analysis in addition to economic analysis. Moyo et al. (2009) proposed the use of a tri-modal agrarian structure. This structure categorizes the results of the reform based on such variables as land size (adjusted to reflect differences in quality and the agro-ecological potential of the land), tenure system that the land was under, control of land holdings, access to markets (support), the class structure of the inhabitants, technical capacity, the crop produced and organisation of production. Analysis of these variables enabled the classification of the beneficiaries into peasants, middle to medium capitalists and large-to-corporate capitalist farmers. This is extremely important for the formation of cooperatives according to Chayanov's theory of peasant cooperatives (discussed in Chapter One and Chapter Two).

The overall national agrarian structure is now dominated by peasants (on less than ten hectares) who are settled on 73% of Zimbabwe's arable land, followed by small to middle capitalists (30 to 150 hectares) settled on 9% of the arable land. The level of land quality/quantity, off-farm income, class, influence, gender and age structures differ throughout peasant farms, which indicates differentiated control of the land markets (Moyo, 2004; Moyo et al., 2014). In general, land seekers in the A2 relied on associational brokering. In the A1, seekers used a mixture of participation (in the occupation movement) and negotiations with local land authorities. However, participation in land occupations, engagement with local land authorities and WVA did not guarantee land access. It took a lot of dedication and commitment. Some beneficiaries had to spend several nights encamped in the forest, in what became known as 'bases' awaiting re-allocation from District Land Committees (Moyo et al., 2009; Sadomba, 2011, p. 108). This has positive implications on the formation and sustainability of cooperatives because new farmers (ex-combatants) believe in commitment, dedication, and the strength of working together towards the achievement of set objectives.

Approximately 85.9% of the land redistribution occurred between 2000 and 2004, with the remainder taking place from 2005 to 2015 (SMAIAS, 2015, pp. 7–8). The FTLRP aimed at decongesting CAs and to lessen rates of poverty, and as such, most of the land beneficiaries (53.6%) came from the CAs while 29.6% came from the urban areas (SMAIAS, 2015). This has implications on the rate at which farmers can come together and form groups or fashion resistance to exploitation in the agrarian markets. It takes time to establish social relations and strengthen kin, kith-ship and trust given that the allocation process did not consider the origins of the people who were applying for land (Chiweshe, 2011, p. 1; Mafeje, 2003; Murisa, 2009). Thus, the type of cooperatives (and their success) depends on the amount of time the members have known each other.

The size of the families also has a bearing on the type of cooperatives that can be formed. Overall, the households with more than six members per farm beneficiary accounted for more than half of the total beneficiaries (53.9%) (SMAIAS, 2015, p. 14). In a case where cooperatives can accept more than one member per household, the newly resettled farmer's family structure has a huge potential to draw many members that can support the network. Furthermore, the level of education within these households pointed to the fact that 84.3% of those who took part in land resettlement had completed primary education (could read and write) (SMAIAS, 2015, pp. 19–20). This is very important for the flow of information within the cooperatives. A significant proportion of the respondents had been 'previously employed' (40.6%) while a slightly lower number (38.5%) had 'never been employed' before accessing land (SMAIAS, 2015, p. 21). This fact can be interpreted as the ability of the members to understand how formal institutions work which could go a long way in improving management skills and sustainability of the cooperatives.

Ownership and access to land for women has been a contentious issue throughout literature. Women's role in agriculture cannot be overstated as they are virtually at the fore of all farm activities from household production to reproduction. However, a mismatch remains between labour days invested and benefits from farming. By 2014, a few studies estimated the number of women who benefited from the land reform in their own right to be between 12 and 20%

125

(Buka, 2003; SMAIAS, 2015; Utete, 2003). Although women accessed land through other channels such as marriage and family institutions, the above result points to a more massive challenge for the resolution of the gender question which hinges on more gender-equitable landholding structure. Some cooperatives require a member to be a land-holder before they are admitted into the cooperative, hence, low access to land restricts their participation in cooperatives.

Chasing after maize: The development of Zimbabwean agriculture

Throughout Zimbabwe history, several events occurred, affected, and shaped the maize production system. These range from the great depression in the 1930s, the sanctions imposed on Rhodesia in the 1960s after the Unilateral Declaration of Independence (UDI), the liberation war (civil war) in the 1980s, the ESAP in the 1990s and the recent 2000 FTLRP.

Zimbabwe is organised into natural regions which determined the degree of intensity of farming that each region could support depending on the amount of rainfall and temperature. On one end is Natural Region 1 (NRI) which is the wettest, had the highest number of estates and is suitable for specialized and diversified farming (owned by companies and white settlers). The NRII is ideal for intensive farming while NRIII has semi-intensive agriculture. The NRIV had semi-extensive while NRV was mostly in the CAs and is suitable for extensive livestock rearing (Vincent & Thomas, 1961). These classifications affected agricultural production and hence affected agricultural policy, for example in the 1980s, in response to improved government subsidy support, the most substantial increases in maize production occurred in NRI-III (Stack, 1994, pp. 258–262). These zones have been further revised in 2020 to reflect the effect of climate change. The general trend is some areas that were high potential are now downgraded to low potential in rainfall and ideal temperature terms while some that were in NRV have been described as unable to sustain any form of agriculture anymore.

Although maize was already being consumed by Africans when the British colonizers came (maize arrived in Africa through trade with the Portuguese in the 15th century), it quickly became a very profitable crop for the white-settler farmers. They produced the crop for export, especially into the British starch industry. In the settler Rhodesia, from 1906, exports of maize grew at an average of 18.8% per annum until 1932 (because of WWII and the global economic depression) (Masters, 1993). After WWII, production and marketing of the crop improved until the liberation struggle in the 1970s as most rural farmers and some commercial farmers abandoned their farms in fear of or to partake in the armed struggle.

The pattern of production before and after independence has been differentiated. Noteworthy is the increase in the contribution of small-scale agriculture to total agricultural production after 1980, mainly because of the removal of colonial restrictions (Binswanger-Mkhize & Moyo, 2012, p. 47). Massive government subsidies (and better output prices) for maize production before liberalisation doubled the number of small-scale farmers from 5% to 10% in the 1979-1985 period (Eicher, 1995, p. 808; Muir-Leresche & Muchopa, 2006, p. 300). Small-scale production increased from 10% to 40% for maize, 7% to 53% for cotton, and 41% to 53% for groundnuts between 1980 and 1987 seasons, respectively. After ESAP, maize production declined by 25% due to reduced support for maize production coupled with two significant droughts. Agriculture played a significant role in the 1980s Zimbabwe economy by contributing approximately 14% to GDP, and about 40% of the total Zimbabwe exports (Stack & Sukume, 2006, p. 567). Additionally, agriculture was a significant absorber of labour and drove a significant amount of indirect economic activities through its linkages with other sectors, such as the manufacturing industry.

The biggest issue was how the state and all relevant stakeholders should develop and maintain a viable and robust agricultural marketing system. Some scholars (Masters, 1993, p. 239) encouraged the government to open more and allow for private-sector competition into the maize marketing system since it was state-run, just as we saw in the rice marketing system in Japan. Some studies have shown that although maize liberalisation improved maize prices, expanded rural trading, processing, farmer commercialisation and improved grain supply to private millers in urban areas (Muir-Leresche & Muchopa, 2006, p. 307), the increase could not off-set the rising cost of production which overall affected rural farmers (Makamure et al., 2001, p. 8). Maize output in Zimbabwe is also greatly affected by drought especially in the late 1980s and the 1990s with the severest drought year of 1992/3 resulting in as much as 1.8 million tonnes of maize imports (Binswanger-Mkhize & Moyo, 2012, p. 55). Another drought which occurred against the backdrop of the land reform in 2003 also had devastating effects on food imports (and food aid); however, maize imports have stabilised from 2003.

The Agricultural Marketing Authority Act (1967) was established to control and oversee the running of most agricultural boards (Cold Storage Commission, Sugar Industry Board, Cotton Marketing Board, Dairy Marketing Board, Tobacco Marketing Board, Grain Marketing Board – GMB). Of interest to this publication was the establishment and running of the GMB for grain markets (Eicher, 1995, p. 811). The GMB had a highly centralized system which was the only channel for farmers to sell their products to consumers. There were three channels that farmers could use, namely i) through the GMB depots (in rural and urban areas) ii) through to the collection points of GMB or iii) through GMB approved grain buyers. However, all the grain had to pass through the GMB. The approved buyers had to submit all the maize to the GMB who would sell it to the consumers/industry (just as in the case of the Japanese rice industry). No private sales were permitted. This resembled the Japanese rice system, with the only difference being the fact that the MAFF approved private buyers were cooperatives and not individuals (as in the Zimbabwe case).

In terms of pricing, the government oversaw the processes of pricing in the period 1980-1990. Government expenditure in agriculture grew in the first six years after 1980 (Makamure et al., 2001, pp. 11–12) and then fell into perpetual decline during the ESAP era. From the year 2000, due to the withdrawal of the private sector from agricultural finance, government spending started to increase (see more detail on page 141).

The Grain Marketing Act

The first set of regulations in the grain markets were introduced when the Grain Marketing Board (GMB) was established in 1931. The Grain Marketing Act of 1967 solidified state control of grain markets and instituted the GMB as the main channel of grain transactions (including soybeans, wheat, sunflower, millet, sorghum, rice, and coffee). Although the board was formed as a temporary measure to whither the harmful effects of the great depression, it stayed until present-day Zimbabwe (Eicher, 1995; Mudege, 2005, p. 79). It sought to keep prices of grains (mainly maize) artificially higher than the world prices and hence shielded the local farmers from the depression, just as in the case of Japan during the Rice Control Act. However, the farmers in this period were predominantly white large-scale farmers who had huge state-support through finance (subsidies), inputs, outputs, and export incentives. The strength of their farmer's organisation enabled them to exert much power in the maize market and even into the Rhodesian economy. The structure of the GMB maintained a dual pricing system, one for large-scale white farms and the other for the black farmers in tribal trust lands. This helped to institutionalize white-settler economic dominance over the black people (Bratton, 1987, pp. 181–182).

Since the GMB controlled the pricing system, where the price set would be just be high enough to incentivize the farmers to sell to the GMB while at the same time, not too low to proliferate the black/parallel market (Herbst, 1988). Upon the attornment of independence in 1980, the new socialist government continued with the policy and sought to bring more small-scale farmers into the marketing system through the establishment of new seasonal grain buying depots (Eicher, 1995, p. 812). This process was almost similar to the Japanese post-war food control system. It wanted to expand the GMB network to include rural maize markets as a way of fixing the historic racial and social injustices of the colonial era. However, the efforts were mainly affected by the droughts in 1983-1984 and 1987, and the economy stagnated. In addition to increasing balance of payments deficits and foreign currency shortages, the drought and economic stagnation drove the state to eventually adopt and implement the disastrous ESAP in the early 1990s.

The large-scale farmers drove the 1960s boom in production, but the small-scale farmers took centre stage after independence. The trajectory taken by the agricultural production and marketing over the last six decades can be pinned on the behaviour of the GMB pricing system set within the Grain Marketing Act. As with the Japanese rice production system, the GMB officials negotiated with farmer representatives (farmer unions) when setting up prices and in other policy-related issues. It is important to note that these farmer unions were not cooperatives but were mainly dominated by large-scale farmers, and hence primarily representing their interest (Masters, 1993, p. 232; Moyo, 2000). The prices were usually set towards the start of the harvesting period; hence, farmers got into maize production cycle with inadequate output price information. Before 1980, the price of maize output was a function of the cost of production; however, from 1980 onwards, it was based on the freight-on-board (FOB) of a tonne of maize. By 1990, about 66 depots (an increase from 34 in 1980) were fully operational with about just under half being in the rural areas. However, private sales and transportation of grains were still unlawful; this meant the pre-independence GMB structure remained in place, industrial millers had to buy from the GMB, and small-scale and large-scale producers had to sell to the GMB.

The GMB employed the pan-territorial and pan-seasonal pricing system in which it bought grain at the same price across the whole country and throughout the year, respectively. Again, this closely resembled the Japanese food control system in which rice had no quality standards and was purchased at the same price. During the era of the regulated market, price setting was done by the government through the GMB monopoly which consulted farmer organisations such as white-settler-farmer-dominated Commercial Farmer's Union – CFU (usually with little small-scale representation). From the government perspective; the rationale was to protect those farmers in highly productive areas from lower prices immediately after harvest (Muir & Takavarasha, 1989, p. 111), and also protect the working class in urban areas for social and political reasons (Herbst, 1988; Muir-Leresche & Muchopa, 2006). However, the fact that the net sellers of grain were the small-scale producers meant that they

130

were taxed, while large-scale farmers who were producing other high-value crops and livestock were subsidized in purchasing this maize.

Here, we see the disadvantage of the poor organisation of the farmers. If they had proper representation in the decision-making echelons, as happened in the case of Japan, they could have negotiated for a mutually benefiting pricing model. In 1988, the producer-surplus loss for small-holders accounted for 17% of their produce; this means they were being paid 17% less than they should have been under non-pan-territorial pricing system. At the same time, the system was subsidizing the large-scale farmers by 9% (Masters, 1993; Muir-Leresche & Muchopa, 2006, p. 302). Pan-territorial pricing affected farm-gate pricing and consumer prices which encouraged parallel market activities as intermediaries found it profitable to buy in the CA and sell to millers in urban areas. If a strong cooperative had been in place, things would have panned out differently.

Liberalization of agricultural markets

Two broad instances of maize markets deregulation occurred in Zimbabwe since independence – during the ESAP and later during dollarization period. ESAP's main agenda was to reduce the amount of government intervention in the whole economy by cutting down the number of ministries and the number of resources that went to ministerial activities. For agriculture, liberalization was interpreted as a deliberate reduction of the government's role in production, distribution and marketing of agricultural inputs and commodities as well as their role in guaranteeing output markets. The GMB maintained a monopoly in purchase, sale, and exports of maize grain during regulation periods. In these times, it controlled cross border trade, private movement across district and zonal boundaries; controlled special supply of grains to the processing industry (Poulton et al., 2002).

Before ESAP and deregulation, government intervention was justified based on i) ensuring that farmers got fair prices, ii) ensuring that urban consumers got cheap priced food, iii) maintenance of food security and grain reserves, iv) taxation of the agricultural produce. Makamure et al. (2001) concluded that liberalization had worsened the plight of farmers, especially for less-tradable crop producing farmers (maize, sweet potatoes). This then affected household food security. This effect can be explained by the fact that small-holder farmers in the CAs did not own private land title (required by banking institutions to access loans); hence, there was no way that they could have benefited from the liberalization.

Liberalization eroded the viability of farming (inflation, interest rates, taxes) because the key stakeholders in the agricultural marketing system were not consulted during the formulation of the policy (Makamure et al., 2001). Such interest groups included farmer and producer organisations (Zimbabwe Farmers Union – ZFU, Commercial Farmers Union – CFU, Indigenous Commercial Farmers Union – ICFU, Cooperatives), agro-industrialists, and individual farmers. In Zimbabwe, large-scale farmers have their associations different from that of the small-scale farmers with little to no inter-association networking for a common goal.

The liberalization of the marketing system also involved a few other policy incentive instruments such as the Export Retention Scheme (ERS), which encouraged exporters to retain a specific proportion (up to 30% depending on the agricultural sector) of the exports in foreign currency (Murisa, 2009, p. 53; Rusike & Sukume, 2006, p. 287). Also, the Open General Import License Scheme (OGILS) enabled free importation of several commodities. The Export Support Facility for funding imports of raw materials for export-oriented production was another incentive but most importantly, the devaluation of the Zimbabwe dollar made exports to be valuable and worked as an excellent incentive for exporters. Unfortunately, many small-scale farmers had low access to the land title (required by banks), and hence they could not access funding to sell their goods on the international markets. Thus, liberalisation benefited LSCF and not my grandfather in the communal areas. The greatest mistake that the ESAP policy had was to assume that all market participants had equal opportunities. In terms of marketing, the smallholder farmers were considerably affected when the GMB removed its temporary collection depots which were close to their places of agricultural production, and this increased their cost of production (transaction costs). My grandfather used to talk about this

132

with vivid sadness. By 1996, there were no GMB collection points in the rural areas which virtually back-rolled the gains of the 1980s (Makamure et al., 2001, p. 20).

Zimbabwe Agricultural Commodity Exchange

As prospects of growth from the ESAP become deceptive, the Zimbabwe Agricultural Commodity Exchange (ZIMACE) facility emerged in 1994 through private sector and state negotiations to try and provide farmers with an alternative to the unfavourable markets (Muir-Leresche & Muchopa, 2006, p. 316). Producers would get bids and offers for their produce from buyers or brokers in attendance, reducing the gap between supply and demand (Makamure, Jowa, & Muzuva, 2001, p. 44). The platform grew because it gave farmers some form of security and transparency as they were trading through legally binding contracts. Maize and other grains such as wheat and soybean formed the bulk of goods traded on this platform. The traded volumes increase yearly by an average of 35% between 1994 and 1996 (Makamure et al., 2001, p. 45). However, such platforms, although functional, could not accommodate all producers from all sectors and membership mainly consisted of Delta beverages, Olivine Industries, large-scale millers and LSCFs. The number of intermediaries who bought maize from the CA and sold to agribusinesses through the ZIMACE platform rose.

The small-scale farmers only benefited from ZIMACE through accessing base price information for their commodity. The minimum transaction volume on the platform was five tonnes, hence discouraging small-scale producers whose output was lower than this threshold (Poulton et al., 2002, p. 42). A group of farmers in Gokwe had formed an organisation to collaborate efforts when selling on the platform. In this situation, selected farmer group leaders can represent the farmers in negotiations. This would also reduce the transaction costs of negotiation on the part of brokers, producers, and the government. This gives scope for the cooperative model because it reveals that institutional order at grassroots level improves structural organisation necessary for agricultural development.

Maize policy after the FTLRP

After the implementation of the FTLRP, the government took control of the maize markets through the Statutory Instrument 235A of 2001 (Muir-Leresche & Muchopa, 2006, pp. 301–316). The GMB received subsidies, bought maize at controlled prices, and sold it to the millers at subsidized rates. This affected profitability of the maize producers and just over half of the A1 (53%) and A2 (58%) farmers utilised the GMB marketing channel (Binswanger-Mkhize & Moyo, 2012, p. 65). The policy soon after the reform punished the farmers and protected the urban consumers, and this is in stark contrast to the rice policy in Japan where prices are still kept artificially higher for the benefit of the grain producers.

After dollarization in 2009, maize markets were liberalised, which saw the increase in the importation of cheaply produced grains from around the region. Zimbabwe was a signatory of the World Trade Organisation (WTO) (just like Japan) they had to liberalize their agricultural sector in line with GATT guidelines. While this improved access to grains, it negatively affected local producers as they had to compete with cheaper processed GMO maize from South Africa. Although the markets were liberalised, GMB still played a significant role in the market primarily through the maintenance of the strategic grain reserve (290,000 tonnes). One of the latest and most exciting developments in the maize policy is the Targeted Command Agriculture Programme or better known as Command Agriculture. It is a Zimbabwe government-led Special Maize Import Substitution 'contract-farming' scheme for large and small-scale farmers using domestic finance capital resources. It brings state, private sector and farmers together to produce food (see more in Mazwi et al., 2019, pp. 6–9). Before such a program, the private sector was involved in maize markets through complex supply chain interlinkages (small-scale agricultural producers, traders and millers); however, for the first time, their participation has been observed in financing the production of maize under 'contract-farming-type' arrangements. The program is, however, marred with corruption scandals, threatening its sustainability.

The Comprehensive Agricultural Policy Framework

Zimbabwe has always struggled to have a comprehensive agricultural development policy since independence. In cases where good policies were developed, responsible ministries have faced challenges in implementing the policies due to inadequate budget allocations, incompetence and sometimes corruption. As noted by Poulton et al. (2002), meaningful attempts to form an agricultural policy framework started with ESAP in the 1990s. Indeed, the Comprehensive Agricultural Policy framework (CAPF) can be equated to Japan's Agricultural Basic Law because it also details the vision, goals, objectives, and detailed development policy strategies. CAPF (covering 2012-2032) is a revised version of the 1994 Zimbabwe Agricultural Policy framework (1995-2020).

The CAPF seeks to organize all institutions and concerned stakeholders (including farmer organisations) in agriculture so as to increase agricultural production, income, and employment as well as to increase contribution of agriculture to GDP. While the Japanese version sought to maintain high levels of agricultural production to achieve food sovereignty and promote rural living, the CAPF seeks to increase agricultural production to achieve food security. It also seeks to develop these rural areas by coordinating input, production, extension, and output marketing activities. However, while the Japanese version was clear in terms of how to deal with the land tenure issues, the CAPF is mute about how land tenure issues should be handled.

Tools of peasant control; A dive into the cooperative society law of Zimbabwe

As we have already discussed (Chapter Three), the state can be manipulated by the neoliberal markets and the wishes of the markets is often engraved in the so-called development policies that the state proposes, adopts, and implements. The Cooperatives Societies Act 1996 (CSA henceforth) has not significantly changed since 1990, although minor amendments were done in 1996 and 2001. In the sections that follow, I provide some of the key issues that have been overtaken by events and necessitates the revision of the Act.

135

Formation, registration, and functions of the cooperative

The minister, in their capacity, through the registrar or the CSA, had too much power within the movement. They have the final decision-making power, which is detrimental to the cooperative (especially within neoliberal and corrupt political environments). For example, Section 3 of the CSA gives the state power to force their ways into cooperative meetings. Section 23 gives the state a front seat during the establishment of cooperative by-laws; Section 35 gives the state power to audit the cooperative movement. The national cooperative association should be given the power to decide who gets to audit their accounts.

On the other hand, the CSA is an old instrument that has not only been overtaken by time (the FTLRP, for example) but still covers eight different economic sectors under one structural framework. Thus, Section 6 needs to be revised to allow for the formation of individual cooperative for each economic sector.

Membership, organization, and management of cooperatives

The cooperative movement faces several challenges (see also Chapter Seven), especially in terms of management. So how does the legislative tool (CSA) improve or worsen the challenges faced by the cooperative movement?

Qualifications for membership of management committee (Section 55) – The variables used to qualify people who can be part of the management committee (for example, they must be members of the same cooperative, never been convicted and should not have outstanding loans) are necessary. However, they should consider academic qualifications and level of skills as well. Also, I suggest the use of the farmer certification to select candidates for the management committee (see more in Chapter Eight). Section 56 states that a cooperative management committee member shall vacate their offices without fail if they do not attend three consecutive meetings of their cooperative. These conditions are not being met; hence, measures to ensure that committee members adhere to this should be placed. Again, the National Federation should enforce this and not the ministry.

Functions of management committees (Section 57) – In essence, the management committee has the power to steer the cooperative to success or failure. Thus, the quality of their skills, education and commitment can never be overstated. These are supposed to meet as much as possible, and half of the committee present constitute a quorum. The problem lies in the clause that says none of these members may receive remuneration for their duties as members of the management committee. Committee members' motivation was a huge factor in the success or failure of the cooperatives. In light of current economic challenges, farmers who carry out cooperative business needs higher levels of motivation to forgo their own private lives and attend to cooperative business.

Duties of the Chairman, vice chairman, secretary, treasurer, and manager (Section 61) – These roles require not only skill but soberness and incorruptibility. Trained personal or holders of the farmers' certificate should fill these roles. Because of low skills and levels of education, some of the elected chairmen did not understand the importance of following procedures as highlighted by one case in which the secretary signed accounting books on behalf of the chairmen, manager, and treasurer. In addition to signing statements on behalf of the manager, secretaries were writing manager/chairman or treasurer reports. This defeated the rational of accountability sought by report writing. Furthermore, the reports are supposed to be sent to the registrar undermining the independence of the cooperatives. I believe a capacitated national federation or apex organisations should receive these reports rather than the government.

Apex and national cooperative organizations
This sub-section contains provisions for the establishment of the highest cooperative organ that can represent the cooperative movement at the national and international levels. It is in this part of the act that the government has maintained a stronghold on the cooperative movement. Cooperative self-funding is one of the fundamental ways of attaining self-sustainability and independence. Thus, a stronger legislature that gives more freedom for the cooperative movement at the national level to be in control of its

137

resources and decisions is necessary. This is particularly true in the current situation where the government is relatively centralised and corrupt.

Formation of the federation (Section 89) – All registered apex organisations, through the approval of the ministry, are allowed to form a federation. They need to have clear objectives, by-laws, structures (chairpersons and other officers) as well as written methods of winding-up the federation when it becomes necessary. Since this is the highest representation for the movement at the national level, the relationship of the cooperative movement and the state/government vastly depends on chairman and vice-chairman positions in the federation. This body is responsible for coordinating economic and other socio-economic plans and submitting them to the ministry. Ideally, this body should be in charge of all things that concern the cooperative, that is to say, the federation should supervise most of the task currently superintended by the ministry through the registrar. However, if the state does this, to a greater extent, it would have surrendered its hegemony on the rural socio-economy. Governments are unforthcoming in this respect (Wedig & Wiegratz, 2018). The federation's financial health reflects on the health of the whole movement. It is supposed to compile information that enables farmers to be knowledgeable in terms of protecting themselves from any social, political, and economic problems. The federation should be able to produce or facilitate loans for the members. These functions are not being practised currently.

Establishment and objectives of a central fund (Section 91) – This establishes a national cooperative reserve fund backed by Apex organisations, member donations and any other monies due. One exciting aspect of the central fund is the composition of its committee. The ministry appoints two members as the representatives of the government; it will also choose three more members from a list submitted by the federation. By doing this, the state effectively has control over the cooperative fund. The central fund is responsible for getting funds for activities such as education and training, auditing of books, research, and development of the cooperative movement. The committee's responsibilities have little to do with policy and legal issues so naturally, the state should not be involved too much in its formation and maintenance. In a case where the government no longer prioritise the cooperative movement, and where the state still controls a central cooperative fund, disaster and doom is the most probable result. I cannot stress this enough; the National Federation needs to be appropriately reconstituted. In section 93, the act stipulates that 5% of the annual proceeds from cooperative activities goes to the central cooperative fund. However, it is mute in terms of how the secondary, apex organisations, and the federation itself, are supposed to be funded.

The State-Market interplay in Zimbabwe: Sites of peasant exploitation

The relationship between the government and the cooperative movement is legally derived from the Cooperative Societies Act (1909 revised in 1922, 1956, 1990 and 1996) and other related acts (Collective Investment Schemes Act, Companies Act, Grain Marketing Act, Rural District Councils Act, Small Enterprises Development Corporation Act and Unlawful Organisations Act to mention just a few) managed by other line ministries such as the Ministry of Lands, the Ministry of Local Government. The Cooperative Societies Act gave overarching power to the government to virtually control everything (see Chapter Six, page 240).

The role of Ministry of Agriculture, Mechanization & Irrigation Development

The administration of the Act has moved from one-line ministry to the other ever since 1980, and at one point was a separate ministry on its own. This greatly affected continuity and some information were lost during crossing over. As shall be revealed through interview data in Chapter Six, the cooperative register has been lost because of this issue. The Ministry of Lands Agriculture and Rural Resettlement, as well as the Ministry of Cooperative Development and Women affairs, administered the Act in most cases. According to SMECD (2017), the government's mandate is to:

- Encourage the formation of societies economy and to promote their efficiency.
- Develop legal as well as regulatory frameworks and implement policies.
- Develop, promote, and coordinate cooperatives as well as monitoring finance schemes.
- Provide training of management, committees, officers, members, and staff of societies.
- Provide business consultancy services and infrastructural services.
- Conduct R&D to improve opportunity environment.
- Maintain databases, monitor activities of the societies, administer & develop funds.
- Recognize and appraise prohibitive laws and regulations.
- Provide platforms that enable high technological adoption rates for the cooperatives.

The heavy hand of the state is evident from these functions that the ministry was responsible for. It virtually controlled all the aspects of the cooperative, including the setting up and voting in of the national federation. In 1989, the World Bank reported that the Cooperative Societies Act was under review to factor in the recommendation from the Ministry of Cooperatives which advocated for the focus to be on the cooperative interest instead of the government interest (World Bank, 1989). Some of the proposed changes included i) the proper adherence to the cooperative principles through the development of a self-reliant cooperative movement, ii) minimum government role in the supervision of cooperatives (ideally, the government should just regulate and promote through friendly policies and not necessarily supervise), iii) introduction of half cooperatives (that can mean a cooperative that is waiting to become a full cooperative) termed pre-cooperative stage, iv) the establishment of a central cooperative fund to support cooperative programs like education, training and audits), v) formation of a Cooperative Tribunal for the settlement of disputes vii) empower the registrar to act in 'extraordinary' circumstances to prevent mismanagement of cooperatives (MYDEC, 2005; World

140

Bank, 1989, p. 8). The registrar was supposed to get more power to revoke any decisions made by a cooperative (democratically or otherwise). The WB supported this because they believed that the cooperative was too young to be given the wings to fly on its own without the help of the government. While this makes sense, I maintain that the cooperatives (through the national and Apex organisations) should take over most of the functions of the registrar that have to do with direct contact with cooperative members.

Agricultural financing in Zimbabwe

Before 1980, the discriminatory dual policy in agriculture ensured that both public and private sector agricultural services (provision of research, advice, credit, and transport) got to the commercial farming sector first, before they could think of the black farmers in the *tribal trust lands*. The Zimbabwe agricultural policy framework has changed several times since independence, mostly in line with the thrust of the national economic production model adopted by the government. In the first formative stages, the country's agricultural finance policy had considerably higher allocations and financing directly from the state to the farming sector. However, as highlighted, this favoured large-scale farmers, even though support to the small-scale increased after independence.

Loans and grants to the small-scale increased from 2% of the total loans/grants from the Agricultural Finance Corporation (AFC, now Agribank) in 1979/80 agricultural seasons, to 17% and benefited about 8.3% of the peasants by 1987 (Bratton, 1987, p. 194; Weiner, 1988, p. 481). Indeed, access to input subsidy programs, infrastructural programs and extension programs for the small-scale sector increased as the government tried to incorporate it into the agricultural marketing sector. The increase in the number of peasants using fertilisers as well as the doubling of maize production corresponded to the increase in the number of peasants accessing credit in the period 1980 (18,000 accessed credit) to 1986 (77,000) (Eicher, 1995, pp. 805–810). If the peasants had structural order, they would have resisted this.

The introduction of ESAP in the 1990s then concluded the once progressive state-funded agricultural policy as it withdrew its dynamic activities in the sector. It reduced public spending in subsidies, infrastructure, and extension, leaving farmers at the mercy of the markets (see page 112). In addition to the dictates of ESAP, the AFC decided to reduce the number of peasants receiving the credit from 77,000 in 1986 to 30,000 in 1990 because of i) managerial and loan supervision issues, ii) droughts and iii) the rate of loan delinquency. In 1994, white-settler farmers were still getting over 80% of the loans from AFC.

Other forms of agricultural financing that emerged post-ESAP, which depended on free-market usury included contract farming, out-grower system and micro-financing (Mazwi & Muchetu, 2015; Moyo et al., 2014). Most of these financing methods focused on non-food crop production (cash crops like tobacco, sugarcane and cotton) which threatened food security except for the new command agriculture program (2018-19) which focused on maize (Mazwi et al., 2019). Therefore, the role of the private sector in the development of the maize grain has been limited only to maize breeding, seed distribution and marketing of high-yielding varieties. Eicher (1995, p. 811) categorically states that during the boom in agricultural production in the 1980s, the private sector did not carry out enough research and that the donor had not fostered robust farming support groups. He argued that the state had taken it upon itself to fund the small-scale sector.

The state reintroduced subsidies and input schemes for food crops such as the presidential maize input scheme during and towards the end of the ESAP. Rural financing has taken two broad approaches after the FTLRP; formal and informal, targeting the two distinct small (CA and A1) and large-scale farming sectors (A2 and LSCFs) respectively. Loans to the small-scale have usually been short term and have accessed more informal sources such as *chimbadzo* or micro-finance schemes. Before the FTLRP, LSCFs used to access both government and private sector sources (Leasing Company of Zimbabwe and Scotfin), but there were high rates of capital flight after the land reform (Zumbika, 2006, p. 342). From 2000-2008, the government (through the AgriBank), had no other choice but to support both small and large-scale farmers. The post-2000 government's agricultural finance policy either used i) direct injections into farming from the national budget through the Ministry of Agriculture or ii) Reserve Bank of Zimbabwe (RBZ) quasi-fiscal operations as agricultural finance channels. The RBZ directly provided cheaper credit to farmers and agribusinesses, instead of through the Ministry of Agriculture (Moyo et al., 2014, p. 25; Zumbika, 2006, pp. 343–345). These allocations declined from 2000 onwards until the start of the dollarization era.

About 80% of the farmers relied on own-savings (from agricultural production), remittances and non-farm income to finance their day-to-day agricultural activities. With the dollarization of the economy in 2009, many lost their life savings. After 2009, the country went back again to the policy of cutting public spending and private sector credit facilities (Figure 4.1). Although this policy managed to revitalize the agricultural sector, it was again focused on cash crops (cotton and tobacco) through contract farming.

Figure 4.1: Lending to Agriculture from Zimbabwe commercial banks (USD)

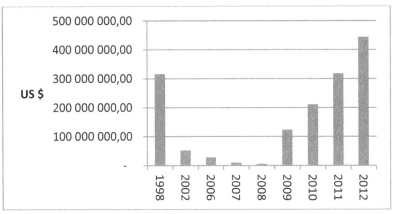

Source: Moyo, Chambati, & Siziba (2014)

In 2012, the government formulated a new 20-year agricultural finance policy framework. The framework centred on three issues; improving the financing of agriculture, increase contract farming to enhance access to inputs and output marketing, and improving equity financing in agriculture. The government, therefore, aimed to increase access to credit for the farmers by establishing an agricultural

143

fund, negotiating with banks to enable them to set aside an agreed percentage of their reserves to agriculture, reduce lending cost and interest rate among other things. Firms were to be incentivised to engage in contract farming through the creation of a regulatory environment for the mutually beneficial contract arrangements. Lastly, the framework realises the need to incorporate international private/individual finance by facilitating and allowing joint agribusiness ventures between local land beneficiaries and international entities (including former farm owners). The most interesting was the need to facilitate the creation of a conducive environment for the development and maintenance of rural savings organisations, better known as SACCOS (MAMID, 2012, pp. 15–16). However, the framework has faced implementation challenges.

Twenty+20 years of interruption: The current condition of the national movement

Cooperatives in Zimbabwe started in the agricultural sector. Other sectors of the economy, which wanted to progress, also utilised the cooperative model as they formed their cooperatives. As it stands, there are about eight sectors or types of cooperatives in Zimbabwe. These are in *i)* mining, *ii)* agriculture, *iii)* tourism, *iv)* housing, *v)* savings & credit, *vi)* transport and communication, *vii)* fishing and *viii)* arts & crafts (Makochekamwa, 2015, pp. 2–5; SMECD, 2017). However, even though all these types of cooperatives require different types of the legislature, management, and monitoring models, in Zimbabwe, they are all administered by one legal instrument – the Cooperatives Societies Act (1996).

Ideology and Institutional structure of Zimbabwe Cooperatives
The cooperative movement in Zimbabwe was and is still structured along the British-Indian type of cooperatives, where the government has overarching power within the movement (Schwettmann, 2000, pp. 4–5). In 1980, due to the socialist influence of Russia and China, the movement adopted socialist ideology which resulted in more focus on equitable distribution of wealth and

144

productive capacity throughout its members. This explains the rise in collectives during the first decade post-independence. Contradictions in these two approaches are still present in the current movement although it has leaned more towards the International Cooperative Alliance (ICA) ideology (which is more neoliberal than the Russian and Chinese cooperative movement). Most cooperatives have generally accepted the ICA ideology because the world is predominately capitalist, and the movement needs to find whatever ways possible to survive, sometimes at the cost of the initial ideologies. For example, scholars such as Ortmann & King (2007) and Cook & Buress (2009) highlight that more cooperatives need to convert into shareholder-owned companies to survive the neo-liberal onslaught.

As already discussed earlier (page 139), cooperatives started in the pre-independence era, where they represented an extension of the government arm into the rural area. It can be argued that traditional African norms (*ubuntu*) and cultural norms played a significant role in the way leaders of cooperatives carried their duties as leaders of such socialist organisations (Muchetu, 2019b, p. 27; Samkange, 1980). Although the movement developed urgency itself, it was always under the hubristic arm of the state, which is ever-present in their day-to-day activities. The Zimbabwe cooperative movement has never completely separated its philosophy from that of the state.

Aligning socialist ideology to British-Indian based cooperatives has often led to ideological confusion. This lack of an independent ideology distinct from the state is illustrated by the reduction in socio-economic activities of the movement each time the state reduced its activities (e.g., ESAP era). It proved that the cooperative movement was not based on a strong ideology but based on farmers who were in the movement to obtain free resources offered by the government. If the movement had an ideology of its own which was well understood by the members, the initial support by the state would have kick-started its development, and it would have survived the absence of government funding and support during the 1990-2000 economic reforms.

To have a clear ideology and structure within a movement is indeed a virtue. Lack of a sense of identity, belonging, and ownership of the means of production in the movement can reduce the potential of the cooperative. It can also be argued that there were two different movements in the history of Zimbabwe cooperatives, each with different sets of trajectories and ideologies. The white-settler cooperative movement was formed and controlled by the white farmers in the 1900s. Just as we saw in the case of Japan, the white-controlled cooperatives had power and influence in the Rhodesian government and influenced the colonial economy. It is also not false to say they spearheaded the formation of the cooperative movement in the black communal farmlands to also benefit from the production in those areas. Given the fact that they had a concise ideology and vision different from that of the state – and consisted of a few thousand farmers – when the colonial state collapsed, they were steadfast, and some are still in existence today (e.g., Windmill and Farmers Corp.). On the other hand, the second movement, whose ideology was steadily dictated and highjacked by the state (the black cooperative movement) could not survive beyond state sponsorship. These were the first to try and foster positive development in the neglected CA and Small-Scale Commercial Farms (SSCF).

The way the current cooperative movement is structured mimics the structure that existed in colonial-era despite several bouts of geopolitical and socio-economic reforms that have occurred at the global and local scale. This resistance to change and reform represents one of the biggest challenges in the cooperative movement (see page 148). The Zimbabwe agricultural cooperative structure consists of four levels (*Figure 4.2*); individual members belong to primary level cooperatives across the eight different sectors of the Zimbabwe economy. These may include village-level agricultural cooperatives or ward level cooperatives as in the case of housing cooperatives. Each of these primary level cooperatives belongs to the provincial or union cooperative level.

146

Figure 4.2: Structure of the Zimbabwe Agricultural Cooperatives Movement

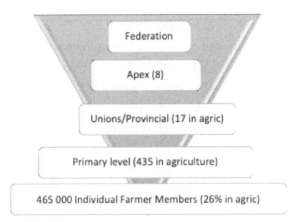

Source: Compiled by author from Makochekamwa (2015) and CACU database (2018)

Each of the economic sectors has an essential Apex cooperative organisation responsible for sector-specific duties. Other countries such as Japan are slowly moving away from a three-tier structure to a two-tier structure of cooperatives by collapsing primary, provincial and Apex bodies into one tier to improve efficiency (Esham et al., 2012, pp. 944–952). At the top of the cooperative structure is the federation, which represents the movement at macro-level dialogue with relevant stakeholders in the cooperative movement (all eight sectors). Power and influence tend to increase moving up the ladder as represented by the inverted triangle (this is true for the pre-reform/1945 JA of Japan). This structure is because of the historically installed government hierarchy in which real decision making about the macro-direction of the movement was in the hands of the federation. This federation's committee was heavily influenced by the state and had several former members of the government serving as federation leaders. This helped the government to maintain a firm grip on the movement and has been a source of dispute in the cooperative reformation debates as shall be explained in later Chapter Six.

Unconducive operating environment: Political interference and corruption

Just like any form of social organisation, the Zimbabwe cooperative movement has its fair share of problems; some are even analogous to the global cooperative movement. These emanate from the contradictions in the neo-liberal socio-economic order, and others are specific to local and national levels. Given that the agricultural movement is governed under the broad Cooperative Societies Act, some of the challenges faced in other sectors (such as the Housing sector) tend to affect the agricultural cooperative movement as well (*Table 4.3*).

Table 4.3: Contemporary local level cooperative challenges in Zimbabwe

Challenge	Most affected sector
Membership contributions and subscriptions	Agriculture and Fishing
A negative perception of contributions	Agriculture and Fishing
Corruption	Housing, Agriculture and Mining
Scattered, weak and uncoordinated cooperatives	Agriculture
Lack of infrastructure	Agriculture
Outdated laws	Agriculture and Housing
Political interference	Agriculture and Housing
Bureaucracy & high license renewal and registration costs	Housing and Fishing
Access to finance and raw materials	Agriculture, Fishing and Arts & Crafts
Marketing	Fishing and Arts & Crafts
Theft and vandalism of property	Agriculture and Fishing
High import taxes and duties	Transport & Communication

Source: Created by the author based on Makochekamwa (2015)

At the macro level, the operational environment has not been conducive for cooperative growth mainly because of unconducive economic liberalisation, ever-changing socio-economic policies (indigenisation, SMEs development, agrarian reform, economic

148

recovery plans), rising unemployment, gender mainstreaming and declining economic growth (Makochekamwa, 2015, p. iii; MYDEC, 2005, p. 4). Other local-level problems for agricultural cooperatives encompassed risky behaviour (conflict and corruption are hidden); inadequate problem-solving skills, and poor decision-making mechanisms (member views are ignored).

Some general issues that are peculiar to the global cooperative movement were also present in the Zimbabwe movement. Such issues included low member participation, coordination, cooperation as well as low levels of education which hinder information flow within the cooperative. Therefore, members are not actively involved in the actual decision-making. Inter-organisational linkages, e.g., with the Zimbabwe Farmers Union (ZFU), National Farmer Association of Zimbabwe (NFAZ) and Zimbabwe National Farmers Union (ZNFU) or the government institutions such as the Department of Agriculture and Rural Extension (AREX) were also weak.

Conflict of interest is another challenge represented by widespread mistrust between the managers and the committee members. The goals of the managers and the committee members often differ, while in most cases the committee seems to have too much power because they are in charge of too many roles and responsibilities. Record keeping and minutes writing were also among significant constraints due to the lack of appreciation of the importance of such tasks for future reference. Cooperative leaders and farmers do not seem to understand why it is done, which affects the filling of essential documents and accounting systems (Chayanov, 1991, p. 47; Chiweshe, 2011, p. 199). Some cooperatives are not even registered. Some, on the other hand, do not have accounting systems cannot draw up final accounts, and hence, the cooperatives are unable to pay out dividends and patronage bonuses.

Options for cooperative sustainability

Several studies have identified the ability to reform and to allow change as the most excellent means to which cooperatives can survive (see Hairong & Yiyuan, 2013; Iliopoulos, 2017; Ortmann & King, 2007; Prakash, 2003). The movement needs to be dynamic and to transform with the changing environment. However, these

149

different scholars differ in the nature and scope of the change and dynamism that must take place. While others believe that more cooperatives should adopt the business model and do away with unprofitable activities (do less free services) to survive (Ortmann & King, 2007), others think that the movement needs to dig deeper into the principles of the cooperative model to survive and also to resolve the various forms of agrarian contradictions that exist in the rural area (Jossa, 2014).

It is not far from the truth that the cooperative movement has the highest potential to alleviate lives in the Zimbabwe rural areas (Romdhane & Moyo, 2002, p. 1). The movement has vast experiences in the agricultural sector as compared to other forms of organisations; they already had 45 established warehouses for distribution of inputs and 250 collection points of outputs by 1990 (these centres were in the most remote areas of the country close to the farmers which made cashing in of farmers cheques easier). They have provisions for structured government support through organised training and auditing of books as well as the free provision of transportation for those that do cooperative business. The cooperative once had a monopoly over input distribution and commanded bargaining power with input suppliers. Furthermore, they enjoyed preferential access to credit from financial institutions as compared to individual applications. Such characteristics drew increased membership and can thus be recreated for the benefit of the peasants. Formation of cooperatives is motivated by the fact that the government cannot adequately serve the rural areas on its own; it is too expensive for the state and too unprofitable for the private sector (Chiweshe, 2011, p. 4). This gives scope for continued support to the cooperative movement.

Summary and conclusions

I reviewed the functions of the GMB by period, from pre- to post-independence, through to the FTLRP and beyond. I discussed these within the context of the resultant post-reform agrarian structure and its implications on the cooperative movement. I did this to give context to the discussion on the history and the trajectory

150

of the cooperative movement. This chapter gave a concise evaluation of the structure of the Zimbabwean cooperative movement, its ideology and relationship with the farmer members, with the government and the options available for its sustainability from a literature review perspective.

The main lessons we can draw from this discussion is that the cooperative movement in Zimbabwe has and is still heavily monitored by the state through its heavy-handed and outdated Cooperative Societies Law (1996). However, the movement has had limited options but to rely on the state given minimised private sector involvement before and after the land reform. The movement has had to rely on the state as the sole provider of finance since independence. The development of the peasantry seems unprofitable for the private sector unless it is at usury rates. This means if agricultural cooperatives, that begin by addressing financial challenges and encourage self-help (and formation of the cooperative banks that prioritises the farmers), then the problem of agriculture financing would be half solved. I highlighted how the government tried to employ the cooperative model with little success. I argued that the failure came from the fact that the government used the cooperative model as a top-down channel of instructions; the peasants were regarded as uneducated folks who were unable to resolve their contradictions. This massive hubristic stance on rural development was the primary source of failure for the development agents over the past half-century or more. The lesson is that policymakers must rethink and restructure programs to include peasants; not as 'subjects' of development but as equal stakeholders in the development trajectory. The next chapter goes back to Japan to understand the current position of cooperatives and how they are dealing with contemporary global challenges within its advanced-capitalist operating environment.

151

Chapter Five

Field evidence; Current trends and patterns in Japanese Agricultural Cooperative System

Introduction

Up to this point, I have covered the historical and contemporary Japanese agricultural system (see Chapter Three), and that of Zimbabwe (Chapter Four) highlighting comparable realities and policies from literature. Striking resemblances in the two country land reforms, their respective grain marketing policies, and the role that the peasant movements played in shaping these policies. The discussion so far has, from a historical and ideological point of view, enhanced the understanding of how cooperatives play an important part, together with the government and private sector, in developing the rural areas. It is now essential to substantiate these abstractions with empirical evidence collected through fieldwork. The primary approach in this chapter was to let the farmers speak and hence, amplify their voices. I identified different learning experiences from contemporary Japanese movement that will assist in the construction of a new agricultural cooperative model for Zimbabwe (see Chapter Eight).

Characteristics of survey participants in Japan

I carried interviews with one academic scholar, two rural activist farmers, three national JA leaders, and the rest were farmers or cooperative members and local leaders. The names of some interviewees were changed upon their request or in cases of unavailability of written consent (Table 5.1).

Kobayashi Jnr (son to Kobayashi senior) is practising farming in the Sanbu and represents a 'rare' success story of generational farmer succession. Nakamura and Kobayashi Jnr. were taught how to farm organic vegetables by Kobayashi. Michio Tsuda and Tohira Kazuo from Yotsuba cooperative are anti-nuclear, environmental, and rural livelihoods activist in Nose district. They oversee the activities of the political wing of Yotsuba cooperative.

Table 5.1: List of interview participants in Japan

Name	Date and place of interview	Organisation Name	Role or occupation
1. Konomi*			
2. Tohira Kazuo*	March 2017, Nose farm	Nose/Yotsuba cooperative	Farmer
3. Abe			
4. Takajin (on attachment)			
5. Michio Tsuda* (Leader)	March 2017, Yotsuba HQ**		
6. Nakamura			
7. Sato			
8. Suzuki	December 2018, Farmer's homestead in Sanbu Village, Chiba	Sanbu network	Farmer
9. Takahashi			
10. Watanabe			
11. Yamada			
12. Kobayashi			
13. Kobayashi Jnr (the son)			
14. Mr Kawakita*	July 2019, Ryuo JA Green Ohmi offices	JA Green Omi	Leader
15. JA Green Omi leaders			Representative
16. Mr Ken Morishima*	Published interview, March 2017	JA-Zennoh	Representative
17. Mr Higa Nakanaka*	Published interview, September 2019		Managing director
18. Prof. Ishida Maasaki*	University of Ryukoku, July 2017	University of Ryukoku	Retired lecturer, coops scholar

Source: Own study; NB* = These are their real names; the rest are pseudo names; ** = Headquarters

154

The research also utilised two published articles from the national JA communications newsletter. The first was an interview of Mr Ken Morishima on the pressure of the JA local, provincial, and national associations to degenerate. The second was a speech by the managing director of the JA, Mr Higa Nakanaka, which focused on the attitudes of the members to the idea, nature, and progress of the self-reform.

I identified a total of 15 sub-themes from the interviews and FGDs transcripts. These became the codes in which I organised the text in NVIVO software. All the themes were of importance as they told different stories, but the seven listed below seemed to dominate the interviews in Japan. Throughout this chapter, I will continuously refer to the following themes:

- *Cooperative appeal-* Appealing aspect of cooperatives
- *Ideology-* Issues that had to do with the ideology of cooperatives.
- *Information-* Effect of information asymmetries on cooperative activities
- *Management-* Issues that affected the management of cooperatives.
- *Political economy-* Political issues that surround the movement.
- *Challenges-* Challenges faced in the overall cooperative movement.
- *Institutional relationships-* Issues that affect relationships of cooperatives with the members, the government, and the private sector.

There exist structural, management and ideological differences between Yotsuba cooperative, Sanbu cooperative and JA Green Ohmi (JAGO) cooperative (see *Figure* 5.1). The differences in extent of focus of the thematic area between the cooperatives reflect the contemporary issues that they face and how they are grappling to find solutions to them; this will provide rich data for the development of the new cooperative model for Zimbabwe. Information flow and education were particularly important for the JA Green Ohmi. While there were statistically significant differences in terms of how the three groups of farmers mainly spoke about their challenges, ideology and political economy, there was no difference in terms of management issues indicating that the latter were intimately relatable across the groups. The data seems to confirm the cooperatives theory which argues that as a cooperative grow and become more prominent, increased heterogeneity intensifies their challenges (M. Cook, 2018; M. Cook et al., 2009).

Figure 5.1: Matrix code frequency query of themes

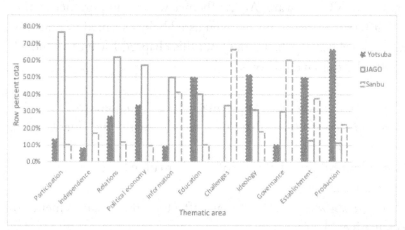

Source: Compiled by the author based on own survey data, 2018-19 (NVivo matrix coding)

I carried a cluster analysis of the coded themes to find out if they were interrelated. Using Pearson's correlation coefficient matrix, I discovered that, naturally, when respondents spoke of challenges, they also had a good idea of the means and steps that would undo the constraints. In the same instance, those that spoke of impediments in the independence of the cooperative also spoke about governance issues.

Figure 5.1 showed that more interviewees from Sanbu network spoke of 'challenges', *Figure 5.2* then reveals that most of these challenges were significantly linked to issues of the flow of information and then to the issues that had to do with the 'independence' and 'relations' of the movement. Another cluster that resulted consisted of politics, ideology and management issues. This result indicates how the management of the cooperative is closely

156

related to 'ideology' and the 'political economy' issues around the cooperative activity.

Figure 5.2: Cluster analysis of themes based on word similarity.

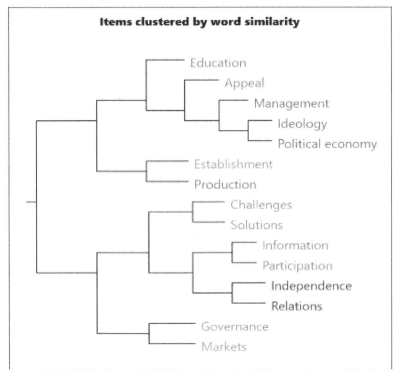

Source: Compiled by the author based on own survey data, 2018-19 (NVivo cluster analysis)

Participation was linkable to education activities and also depended on the reason why members had chosen to join the cooperative ('appeal' in *Figure 5.2*). Finally, issues to do with markets and institutional relationships were discussed within the context of levels of production. In the following sections, I delve more into the themes and try to understand how each affected the Japanese agricultural movement to draw learning points.

Japanese farmers' narratives: A bag of mixed emotions

Dore (2012) wrote extensively on how the cooperative movement of the pre-capitalist Japan was bottom-up with most of the farmers joining the cooperatives freely to benefit from the economies of scale. Ishida (2002a) argued that although farmers found the cooperatives model appealing, the landlords and village heads in those times encouraged people to work in cooperatives to make it easier for them to collect rice taxes. On the other hand, George-Mulgan (2005) argues that the post-war demand for cooperatives was state-driven calling the Japanese government an interventionist state (discussed in detail in Chapter Three, page 91). However, while George-Mulgan's argument is plausible, especially given the fact that the JA had a monopoly in the cooperative sector due to mandatory farmer membership, other more substantial issues also exist. The results of this sub-section provide useful data for the development of the cooperative model for Zimbabwe in Chapter Eight.

Yotsuba cooperative (よつ葉生活協同組合)

The story of Yotsuba cooperative is fascinating. It has two distinct but well-integrated structures. At its base is a socialist organisation that closely resembles a collective, where the central unit of production is the cooperative, and not individual household. The socialist organisation is based on a farm known today as Nose farm. Nose is not only perceived as a place of agricultural production but also as a place of "self-transformation and self-enlightenment". More of the members joined for ideological reasons than they did for increased profits. Mr Ueda established the farm in the 1980s. He was a former municipality officer who got sacked from his job because of his activism aligned with the Chinese way of communism. During these times, Japan had a firm stance against communism which was bankrolled by the USA. The farm was established to provide an income and food for the ideological school that Mr Ueda wanted to open. The members and students who joined the communist movement could not get employment from the state or even from other farmers because they were known as communists (Tsuda interview, March 2017). Deeply entrenched in their ideology was the need to live in harmony with nature, and hence, the reason why they later focused on organic production. Many of the first workers/members at that time were university student activists.

After forty years of existence, the method of recruiting members remains the same. Those farmers that are interested in the way of life of Yotsuba are welcome. As indicated, its philosophy is more inclined towards 'a collective' rather than a cooperative in its purest sense. The Nose farmworkers receive the same monthly salary (¥90,000 as of March 2017) irrespective of their positions and roles on the farm. The cooperative provides for food, accommodation, working tools in addition to the basic salary. Members take turns to prepare breakfast, lunch and dinner, eat together around a huge dining table (discussing the events of the day and preparing for the activities of the following day and thus acting as pseudo daily meetings). After eating, everyone washed their dishes by themselves. Young students from universities are admitted into this farm for an internship so that they experience this way of life, and if they find it interesting, then they may also stay and work on the farm. According to the farm head, 60% of the students always come back after they finish their internship.

> I was born in a prefecture near Tokyo. I did my internship at this farm, and I fell in love with it. I then found a job to work here and have been working for six years now. I fell in love with the kindness of the people and their way of life, which treated everyone with equal respect. I also loved the philosophy of Nose farm and their focus on natural or organic farming. – (Konomi interview, March 2017)
>
> People call me Abe. My hometown is in Takatsu, which is in Osaka prefecture. I started working at one of the Yotsuba offices in my hometown, doing office work. I then met one of the people who would lead me to like, and study [introduced me to] Che Guevara philosophy on how to live a healthy life in a classless, oppression less and free society, this motivated me to want to come and live on the actual farm. (Abe interview, March 2017)

The whole purpose of the farm is to improve the lives of the members; therefore, the profits from the farming operations are invested in more farming units, infrastructure and education (Tohira interview, March 2017). Nose farm has two farms, one in Nose district and another in Hyogo prefecture both producing cattle (over 200 as of March 2017). The farm has 12 permanent members/workers, however, through cooperative labour arrangements with other farmers and cooperative members around Nose district, the farm has on average 21 people that are available for farm tasks every day. These 12 permanent workers also provide labour to the ageing farmers with some labour-intensive tasks like planting, weeding and harvesting.

Yotsuba cooperative was formed from the philosophy of Nose farm to take care of marketing operations when a parent cooperative to which Nose belonged broke up. The parent cooperative wanted to focus more on consumer needs, but Yotsuba maintained their allegiance to the farmers in the Kansai prefecture. That is the reason why most farmers decided to join Yotsuba instead of the consumer-oriented cooperative. However, Yotsuba with a membership of over 40,000 now, is also associated to another consumer cooperative, Kansai Yotsuba Renraku Kai and has close relations with other producer cooperatives such as Hokusetsu Cooperative Farm, and Setoda Farm (with farms located in Kyoto, Nara, Hyogo, Osaka and Shiga prefectures). In addition to wholesome cooperatives joining the Yotsuba group, many individual farmers sell their organic agricultural produce via Yotsuba as well. This represents the second structure in which the individual farmers are recognized as a separate unit of production.

Yotsuba cooperative attracted individual farmers for several other reasons beyond ideology and its focus on producers. Many of the farmers felt that JA had become more concerned with the needs of urban consumers at the expense of the original farming households. Some farmers highlighted that the focus on organic farming attracted them. In contrast, others wanted to access the marketing channels of Yotsuba, which seemed to fetch higher prices than their marketing channels and that of JA.

The farmers themselves want to join Yotsuba because of the prices JA offers are low. Here in Nose, there are around 200 farmers who have joined Yotsuba. JA has recently presented itself as more of a consumer cooperative than a producer cooperative. (Tohira interview, March 2017).

Sanbu Yasai Network (さんぶ野菜ネットワーク)

While the story of Yotsuba was intriguing and more drenched in leftist ideological foundations, the establishment of Sanbu Yasai network (Sanbu vegetable network; Sanbu henceforth) provided many lessons as well. Farmers used toxic materials in the 1950s-1960s in their agricultural production, before those around Chiba area decided that it would be better to reduce the amount of chemicals during agricultural production and charge a small premium for chemical-free vegetables. Eventually, this became part of a movement in Japan in the 1970s, and it was at this time that farmers teamed up to grow organic vegetables. Thus, healthy food was produced using methods that were friendly to the environment while increasing the overall income.

> I can grow vegetables without using any chemical fertiliser, chemical pesticide or herbicide. Reliable and safe vegetables can put a smile on many people's faces. This fact is priceless for me, so my life is better after joining the cooperative. – (Suzuki interview, December 2018)

In the first days, the cooperative was registered under the JA Chiba district structures in 1988. At the same time, it was decided to form a delivery company that would deliver the vegetable right to the customer's doorstep. As the network became more prominent and it realised substantial profits, it became easier for others to join the network. The year 2018 marked the 13th anniversary since branching from the JA in 2005. They decided to leave the JA group because of intra-structural issues with the leaders because they felt that the JA was no longer representing their interest. The JA seemed to be 'stuck in a time loop of the boom years' and that the JA was failing to move with the changing times since it took too long to make decisions (Nakamura interview, December 2018).

Sanbu is a small network of about 61 farmers who produce organic vegetables individually and sell as a cooperative; 81.9% were active contributors, and approximately 49.2% had joined the network less than five years ago indicating growth in membership in recent years. A warehouse in Sanbu village acts as a central hub for vegetable collections and sales through various channels including supermarkets in Tokyo, Chiba and at the Organic farmer's market. The network is strict about the use of genetically modified organisms (GMO) and health of the consumers (see *Picture 5.1* – the writing on the pictures loosely translate to 'we are a health-conscious and GMO-free vegetable producing network'). There are no subscription payments; instead, each member contributes through 20% of the income made from products sold through the cooperative. This management was a refreshing style because the payment of subscriptions is one of the most challenging issues in cooperative management. When the cooperative makes some profits, they redistribute them to the members, and this was a vital source of attraction for the farmers to join the network. Sanbu is not the most prominent network or cooperative in Chiba area, but some other of like-structured networks (some smaller and some bigger) exist.

Picture 5.1: Cooperative warehouse for receiving and dispatching farmer's produce. Source: Field research photos in Sanbu Village, the author, December 2018

The need to grow organic vegetables was thus the most valued aspect of the network, while increased income was the primary motivation to join Sanbu. In this respect, it operates like Yotsuba because it also links farmers around it into various markets. Some farmers scorn the strong ideology of Nose farm, but they bring vegetables because Yotsuba buys them at reasonable prices. Some of the farmers enjoyed the social interactions among themselves, with the consumers as well as with companies (input and service providers). Some extra-marketing technics utilised included farmer-identifying-markers, and sometimes pictures of the producers were displayed in the supermarket (Picture 5.2). This increased farmer's satisfaction with the network and worked as a way of motivating the farmers while increasing the information sharing and better pricing between farmers and the consumers.

I think the interaction between consumers and farmers is the most remarkable aspect of the cooperative; it allows backwards and forwards feedback of information. We sometimes hold meetings with consumers, and sometimes consumers write messages to us. It motivates us to work even more challenging. (Kobayashi Jnr interview, December 2018).

Picture 5.2: Products are identified by farmer in the cooperative stand inside a supermarket.

Source: Field research photos in Sanbu Village, the author, December 2018

I like the fact that I meet new farmers, and we share a lot of *beginner's experiences* and ideas as well as to find creative ways of solving some of the challenges in farming. – (Nakamura interview, December 2018)

I can meet many business associates through the network. By next year, I would have been in the network for 30 years. It was a challenging milestone to achieve, but it is also a rewarding thing for my life. – (Sato interview, December 2018)

On the other hand, several farmers appreciated the amount of new information that the network gathered and relayed to the farmers on time. This information improved their efficiency by reducing transaction costs.

I especially appreciate government information targeted to farmers. The network takes this information and re-packages it in an organised and easily understandable way for us farmers. For example, I learnt about the housing loan scheme for farmers through the cooperative. – (Nakamura interview, December 2018)

Since I belong to the cooperative, I can get information about government programs and subsidies, which is very important for me. I cannot easily access this information by myself. – (Kobayashi Jnr. Interview, December 2018)

Other farmers highlighted that one of the biggest production challenges in Japan was securing output markets. Therefore, joining the cooperative effectively reduced this problem.

I joined the network to never worry about securing markets and hence focus on growing vegetables. So, I concentrate on the production of various vegetables, and they worry about marketing my produce. I also thought that organic vegetables would make more money than conventionally produced vegetables. – (Takahashi interview, December 2018)

Others viewed the cooperative as a method of improving their status in society. What this meant was that being part of the network was viewed in high esteem. As one of the younger farmers who grew up within the structures of the network, and whose father held a decision-making position explained that:

> I am the successor of one of the organic farms, so that is how I joined, by taking over part of the farming that my father was doing. I worked for an insurance company after graduating from school, and then my father introduced me to organic farming. I finally chose to join the network because, in the network, the individual identification sticker on products represents me. So, to me, joining the cooperative enabled me to realise not only economic benefit but also benefits in terms of social interaction. I attend meetings in my capacity as a potential executive of the cooperative because my father used to be an executive before. – (Kobayashi Jnr. Interview, December 2018)

JA Green Ohmi (JA グリーン近江) – JA-GO

JA Green Ohmi (JA-GO) is a cooperative consisting of nine district-level JA cooperatives established on the 1st of October 1994 in Shiga prefecture. The cooperative has various business activities, mainly credit, mutual aid, purchasing and sales, tourism, asset management and welfare of the elderly. It has a total of 23,884 members (8,645 as regular members and the rest are associate members) with a total arable land area of 13,500ha (740ha cropped area) as of 31 July 2019 (JA-GO, 2019). In the last ten years, the number of local-level cooperative members has grown from 5 in 2008 to 131 in 2019 (*Figure 5.3*) with each cooperative having 30-50 households. On average, 70% of the households are rice paddy farmers whose incomes have improved after joining JA-GO (Kawakita interview, July 2019). This is a significant positive aspect of JA-GO. It managed to increase members benefits amid decreases of farmers satisfaction throughout other regional JAs. A Good Agricultural Practice[16] certification was awarded to the cooperative from Germany, showing how JA-GO has been able to stick to cooperative principles and guidelines.

---

[16] This is a set of global certification standards based on food safety, traceability and adherence to environmentally friendly production methods.

Indeed, Japan used the system of certification to its fullest. There are several incentives that certified farmers can enjoy which includes the ability to participate in land markets, ability to run for office in the cooperative leadership, access to subsidized loans, preferential lending and taxation, access to markets that require strict adherence to traceability and integrity among other things. The certification program has several uses including track record of the farmer's productivity over time which helps in credit scoring (OECD, 2009, p. 61). Undeniably, Zimbabwe farmers can utilize a modified version of this.

JA-GO has bank, credit and insurance portfolios which were instrumental in attracting agricultural and non-agricultural members. The banking branch is autonomous in selecting the target of the beneficiaries at the local level but still operates under the cooperative law. Membership participation in their district is almost 100% as is seen throughout other regions in Japan with 'only about 4 out of 131 who sometimes show hostility towards the JA-GO (Kawakita interview, July 2019).

Figure 5.3: Number of local agricultural cooperatives in the JA-GO region.

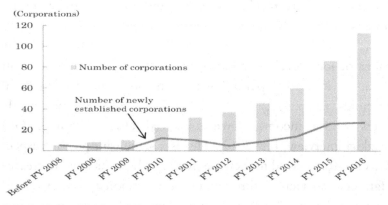

Source: Saito (2018, p. 13)

I spent some time in the JA-Green Omi in Ryuo and participated in activities between JA-GO and a consumer cooperative called *Seikatsu Kurabu* (Life Club) in Kyoto. The consumer cooperative has

166

strong ties with the Ryuo district farmers who are members of the JA Green Ohmi. The farmers in Ryuo were relatively autonomous and were JA members individually, but no formal group existed outside of the JA itself. This revelation was interesting for this book because it made it easy to understand micro-village realities in terms of how farmers perceive the role of JA in contemporary agriculture. Many of the farmers across Japan complain that the JA has neglected the farmers. They highlighted that JA was now more concerned with other economic activities such as insurance, the JA-GO seemed to have maintained a firm focus on increasing the standards of living for the farmers.

Picture 5.1: Participatory field research, planting activities and price negotiation meetings. Source: Field research photos in Ryuo Village, the author, June 2018

In each activity in which the researcher took part, 10-12 JA-GO officials were always taking part. The farmer's field hosted planting activities, while JA-GO office spaces hosted product marketing meetings (see *Picture 5.3*). There seemed to be a mutual understanding between the farmers, the cooperative members and the consumer-cooperative leaders.

Many of the officials who organised and attended the activities were former university students of agriculture. In an interview with a university professor, I learned that the JA used to absorb many of the agriculture graduates. This seemed to persist in JA-GO (Ishida interview, July 2017). In Ryuo, I found that some multi-national companies that had a global monopoly in fertilisers, pesticides and herbicides had to partner with JA-GO to sell their products to the JA-GO farmers (*Picture 5.4*). In doing this, the farmers take part in the value chains through the cooperative, hence benefiting the cooperative members through lower input prices.

Picture 5.4: Round-up herbicide of Monsanto/Bayer in partnership with JA cooperative.
Source: Field research photos in Ryuo Village, the author, May 2018

The three cases are quite distinctive. First, Yotsuba is a consumer cooperative that wants to respect farmers' livelihoods. Nose Farm is a sort of collective cooperative as a part of Yotsuba, which became possible because their activity is capital intensive cattle breeding. Its members are not farmers *per se* because the unit of production is the farm. At the same time, Nose Farm seems to function as a node of local village farmers by providing sales marketing services. Yotsuba tries to bridge the rural and the urban as an Osaka-wide regional cooperative, with an ideological emphasis being placed on the rural. Their behaviour seems to be explained better through Marxist-

Leninist theory of cooperative behaviour.

Second, Sanbu is a village cooperative, though they do not belong to a Green Ohmi-style regional organization nor a national federation. Sanbu's leaders also share the origin of the new left movement as Yotsuba, but it seems that they wanted to become more "realistic". Like small radical trade unions often fight big bureaucratic trade unions, Sanbu's leadership seems to be eager to dismantle the old conservative national JA structure. To this end, they seem to have sought to establish some links with liberals in national politics. Their behaviour seems to focus on creating a robust institutional structure for penetrating liberal markets and making profits; they fit more into the NIE and NCE theory of cooperative behaviour.

Thirdly, JA-GO is a regional cooperative that forms part of the national federation (JA-Zennoh). JA-GO has tried to organise village cooperatives quite successfully under its umbrella. It strives to revitalise the old JA. They do not control but facilitate activities of local village cooperatives (like Sanbu), promote organizational unity in the Ohmi region, and wield some national political influence through the nationwide JA structure. Ideologically, they are neither left nor right; they want to promote the interest of farmers. While they have aspects of both political right and left, their focus changes with respect to what benefits the farmers; thus, they are inclined towards agrarian populism, hence ideally fit into the Chayanovian theory of cooperation.

Importance of strong cooperative ideology

Yotsuba utilised its ideology more strongly in management, production, the flow of information and the general sustainability of the cooperative movement than the other two.

> Nose farm started to provide an income and food for the ideology school because the members and students could not get employment. They were labelled communists. So, the farm sought to educate the young activists in Japan. Explicitly, to teach them about nature, how-to live-in harmony with nature. (Tohira interview, March 2017)

169

Philosophy (a hybrid of Maoist and Che Guevara teachings) leads all the activities at Nose farm. It is visible in the cooperative's approach to organic vegetable production. The philosophy encouraged the free flow of information and treating of all persons as equals during meetings which improved cooperative management over time. There are no superiors during the allocation of farm tasks; in fact, leadership for various activities was rotational and provided opportunities for each member to learn and teach others. The general business model is also influenced by philosophy as it seeks to prioritise the needs of the members. The farm encourages an interactive social life in which workers/members depend on each other and work as a team. They are encouraged to be kind to everyone because everyone is equal in importance. Meals are taken together in a communal dining hall.

> It started from the beginning, the idea of co-living and sharing everything. It started with the students of Ueda san. Staying as one unit helps in teaching people the importance of respecting others and also always thinking about the welfare of others and reduce the need to be alone all the time. People are starting to like how we do things. I see Nose having a positive effect as people are starting to see the wrong side of market-led agriculture, and inorganic production, which is unhealthy and responsible for natural disasters. – (Tohira interview, March 2017)

The farm is a regional base and aims at social reform, resisting adverse effects of globalisation, and nurturing critical consciousness among farmers. A portrait of Che Guevara hangs in the dining hall. One of the things that came to mind was the fact that maybe this collective set-up was some kind of an indoctrination group, teaching people to follow communist ideologies. However, from my survey, I discovered that the Che Guevara portrait had been put by a former member and no one had taken it down ever since even though the member was no longer with Nose farm. The farm teaches about agricultural production philosophy, but if they are also interested in such things as Guevara, they are taught as well (Teramoto informal interview, March 2017). The farm hopes to encourage this type of lifestyle within other farms. At the moment, the Yotsuba head office cooperative staff adopted this structure as well.

In line with the farm objective of "self-transformation and self-enlightenment", a banner with the farm charter is erected to remind members of cooperative vision and goals. Although the charter is important, it is not enforced upon the members as if it is a law but were just guidelines for an ideal society (Tsuda interview, March 2017). Members can choose to adhere to them or to ignore them, and no repercussions will befall them. The charter was made in 1991, and summarised as the following:

Our purpose is to achieve human emancipation through:
1. bringing together as many people as possible,
2. fostering spaces for self-help and equal social relationships,
3. recuperating the art of work, in which people enjoy working together,
4. aiming for the symbiotic relationship between human and nature on a global scale,
5. working and learning together in our community life,
6. not discriminating each other in terms of working ability,
7. criticising each other and avoiding easy collusion and conspiracy.

Source: Observatory Nose/Yotsuba field survey, 2017

Farming for **Sanbu** was more of a guideline or a set of principles for producing food that is safe for the consumers and the environment. The target of philosophical teachings in this particular cooperative is the improvement of products to increase the overall individual income from agricultural activities. One of the founding fathers of the network confirmed this by saying:

There was a movement in Japan in the 1970s about environmental concerns, and it was at this time that I decided to team up with several farmers to grow organic vegetables. This would also serve the purpose of introducing a new product into the horticultural business, which would be sold as organic vegetables for a higher price and hence more

171

profitable [...]. As I became a prominent farmer using this method and realising substantial profits from it, it became easier for others to join the network. (Kobayashi interview, December 2018)

It may be argued that in the purest sense of the term of ideology, Sanbu Yasai network did not have a laid-out philosophy. Their main concern was to produce food for the consumers and be able to increase the profits of the farmers. It was purely a business motivated by profit maximisation. Sanbu represents a set of smaller-scale farmers that engage in collective action to compete well within the capitalist world. They are not fighting it; instead, they want to find ways in which they can benefit and survive within the system. Some of the farmers had extreme, self-proclaimed liberal views on the way that the agricultural system of Japan should be set up. One of the farmers said:

The free trade concept is OK for the cooperative movement. However, many cooperatives say they do not want the (agricultural) sector to liberalise, but I think it is good to have free trade even in fresh vegetables. It should be OK to open the agricultural sector to the private sector and companies to come in, and hence regulations should be discarded. However, at the same time, cooperatives should be allowed to participate in other economic sectors as well (such as insurance and finance) to spread their risk and survive. (Nakamura interview, December 2018)

However, some farmers in the same cooperative held different views about the way on which cooperative philosophy should focus. While older farmers (the founding members) believed that the cooperative was extraordinarily successful and that they were doing the right thing, others, more energetic and younger farmers felt that the cooperative was losing focus. One of the younger farmers highlighted that, the older leaders viewed the cooperative as their private property and hence were slowly ignoring the plight of the farmers by focusing more on the consumers than farmers.

We can work together for a common goal although at times I doubt whether the cooperative improves my life, financial wise that is [...] I think the organisation is losing sight of its objectives because of their need to make profits. There are also times when the need to follow management protocol also affects the overall goals of the cooperative. Also, the purpose of the network needs to be more explicit, as well as its significance. (Watanabe interview, December 2018)

For JA Green Ohmi, because it is still under the strong influence of the JA, it can be argued that ideological basis followed that of the national government. Prof. Ishida explained that:

Agricultural unions had a strong Marxist orientation about them before the end of WWII. So, this was not desired by the government, thus the abolishing of unions and the establishment of less ideologically driven agricultural cooperatives. (Ishida interview, July 2017)

Therefore, the Japanese government policy plays a considerable role in shaping JA-GO philosophy. However, local-level cooperatives can define their paths, which may account for local-level differences in cooperative success. For example, unlike many other JAs across Japan, JA-GO has maintained an ideology in which the individual farmers maintain some advanced level of autonomy, while actively participating in the overall activities of the cooperative. Thus, JA-GO tries to obtain consent from the members through surveys and meetings before implementing any local policies.

[...] I would like to thank the efforts of JA who continue to investigate (look for information about members perspective). Also, Union members should continue the dialogue. This time, we raised a dialogue campaign based on the survey, but we would like to propose a policy for more future dialogues. (Nakanaka interview, September 2019)

I also found that in the case of JA-GO, which has very close ties with the consumer cooperative, pursuit of one clear-cut ideology for

173

a producer may be difficult since they always needed to balance their interest with the consumer cooperative. These two types of cooperative operate on a different ideological wavelength. While all the three discussed cooperative approaches are essential and very interesting, their applicability in developing countries and societies such as that of Zimbabwe would differ immensely. However, I sought to use these insights to construct an ideal ideological path for Zimbabwe (see Chapter Eight).

To conclude, the three cooperatives revealed a lot about the importance of a clearly defined ideology. Yotsuba ideologically focused on a disciplined way of life, Sanbu on the processes of production while the focus of JA-GO was widespread with an inclination towards resolving political-economic contradictions. For Zimbabwean farmers, who care more about increased incomes than about ideology, the focus (at least in the initial stages) should be to provide services otherwise unavailable, most notably financial capital. Thus, an ideology that is 'business-oriented' like Sanbu and yet leaning towards JA-GO's selective use of ideology and politics for the benefit of farmers would be ideal.

Information asymmetries

Hayami & Godo (2005)'s CMS theorem argues that substantial information asymmetries exist in rural areas. Intra-cooperative information asymmetries also play a significant role in undermining cooperative growth and development. From the interviews, Nose farm seemed to have the lowest levels of intra-cooperative information asymmetries attributed to holding of meetings almost daily, during meals and after meals. The overall Yotsuba cooperative held bi-monthly meetings. Given the structural set-up of the cooperative, every individual is regarded as equal, so their input in the cooperative is considered necessary. In doing so, the leaders of the cooperative provide all information about cooperative business activities to every member. This seems simple in theory but arduous in practice. The farmers also share information among themselves and sometimes hold teaching seminars on machinery use and new technologies, especially to the younger aspiring farmers (Konomi

174

interview, March 2017). The leader of the Nose farm pointed out that:

> We have farm meetings every day at the breakfast/lunch table for things such as allocation of duties and to a discussion of unplanned circumstances that arise. Our AGM [Annual General Meeting] for the sharing of the decisions made by the executive is regularly held in March. Customer company representatives are invited to this AGM as well as the local government. Information about the financial position of the cooperative is made [public] in this meeting. (Teramoto interview, March 2017)

Empirical evidence seems to suggest that the structure of Nose farm and that of Yotsuba farm, where there is a strong emphasis on ideology to lead all aspect of farm life, production and marketing, enables it to achieve lowest levels of information asymmetries.

JA-GO also emphasised on the crucial role that the flow of information within the cooperative played. However, because JA-GO was a massive cooperative with over twenty thousand members (within regional assemblage village cooperatives), the flow of information was problematic. For that reason, it was important for the JA staff to carry out as many dialogue activities and opinion surveys as much as possible. Additionally, the farmer's thoughts and opinions were collected through the group approach.

> There is an understanding between the members and the JA-GO [...]. Decision making is a process. The top management/leaders prepare a policy plan in consultation with their members at the local level. In strategy planning, officers and employees will consider drafts based on opinions from union members and users. The meeting of representatives discusses it further before improving it at an annual meeting/conference. We have successes and some failures, of course, but because of the size of the cooperative, we cannot skip any stage. (Kawakita interview, June 2019)

Conversely, among all the issues discussed with the farmers in Sanbu, the quantity and quality of information seemed to be of great concern to the members. While most of them pointed to the fact that they received a significant amount of information about prices of commodities and quantities on demand, they did not have a mechanism for their ideas and information to be communicated to the management team. Thus, information flow in Sanbu Yasai network was uni-directional. The fact that elder members of the network (founding members) that hold relatively senior positions in the cooperative were satisfied with the information flow compounded the problem. For example:

Information such as sales, losses and prices of produce is shared widely. However, information on the rationale of decisions made by the executives is not shared but is kept within the top leadership. (Takahashi interview, December 2018)

There is no problem with the quantity of the information; however, the quality of information flow leaves a lot to be desired. There is some specific information that we require, and we do not get it. (Watanabe interview, 12/2018)

Another group of interviewees were satisfied because they did not expect too much from the cooperative. One farmer pointed to the fact that although the cooperative should be a one-stop-shop for all needs of a farmer, they should be able to access information from various other sources such as the internet and the newspaper without relying on the cooperative.

I get about 60% of the information needed for agricultural production from the cooperative and am satisfied with it. I, however, have got to find the remainder (40%) from other sources. Since I belong to the cooperative, I can get information about government programs and subsidies, which is very important for the farmers. (Kobayashi Jnr. Interview, December 2018)

I am satisfied because I do not expect too much from the network. I especially appreciate information from the government for farmers that comes through the cooperative [...] (Nakamura interview, December 2018)

Therefore, information flow within the cooperative is of great importance, as seen through Yotsuba's ideological based approach and JA Green Ohmi's systematic approach. The data from Sanbu network showed us what can happen if the information does not flow perfectly between management and the members. The model we shall produce should take great care to curb information asymmetries.

Institutional relationship matters

Let us now try to understand the inter- and intra- cooperative differences between the three cooperatives, and how these can apply to develop countries like Zimbabwe. Again, in the context of the CMS framework, the institutional relationships are exceptionally vital for this book because it enables us to understand how cooperatives fit into the community vertex. For Japan, which is predominantly capitalist, strict market forces regulate access to factors of production such as credit, finance and inputs. However, even though the government of Japan has been reducing its support to the rural areas ever since the disbandment of the Rice Control Act (the 1990s), there still exist several subsidies and programs earmarked for the agricultural sector.

For Nose and Yotsuba cooperative, considering their deep socialist socio-political standpoint, most of their relations with the private-owned institutions are governed purely by market forces. Although it has some structures that carry extensive fundraising, it has to borrow from banking institutions from time to time. The leader of the Yotsuba cooperative mentioned that:

Nose and Yotsuba borrow from the commercial banks, and that is how they raise their financial capital. Yotsuba also has some other companies that are also into the finance business, or they organise financing on behalf of Nose. The relationship is that all the products from the farm go to Yotsuba and it sells it through its various channels. For example, the cows that are slaughtered here go to the private

177

company for processing, sales and then later pays us. (Tsuda interview, March 2017)

Depending on the type of animal, livestock at Nose are fed different types of feed and in differentiated frequencies. Some of the feed comes from by-products of the food industry in Osaka such as *sopa* and soybean by-products (Konomi interview, March 2017). That means some of the relations are within the contexts of mutual business relations which does not transcend beyond that. I also observed that some of the cattle feed comes from outside the country. In this sense, the farm and cooperatives depend on foreign agrarian markets or agribusiness to meet livestock feed requirements and hence, the need to maintain good relations with importation companies.

As was highlighted earlier (page 157), a number of the members who work at Nose farm belong to or are ex-graduates from universities. Nose farm accepts university students for internship, and thus, they also developed good relations with educational institutions. They also hope that these relations can help spread their message across to the younger generation. I discovered that some surrounding local cooperative members were also part of the JA Nose district cooperative and that there was some mutual understanding between Yotsuba cooperative and the JA Nose district structures. In the same sense, there also existed other cooperatives around the area, and Yotsuba had some relationships or some established ground for meeting and discussing potential areas of conflict.

Furthermore, just as does many other social organisations (even the *Yakuza* does so), Yotsuba provides a helping hand in times of natural disasters such as typhoons, earthquakes and floods as highlighted by Tohira:

> [...] some of the farmers belong to the JA and some to other cooperatives, and they want to join us, so sometimes the relationship is through the farmers themselves. We have to develop relations for the benefit of the farmers. Another one is called Farmer's Union, which belongs to the communist party and has very few memberships. Their

178

prices are also above JA prices for buying rice. […] We also help when there are national natural disasters. For example, during the earthquake times, Yotsuba helped with money, food and this one time, we had to go there ourselves. (Tohira interview, March 2017)

What we can learn from this information is that there is a need to have a broader social and business network for a cooperative to develop. While Yotsuba has a progressive philosophy, it acknowledges that the rest of the world may not share the same ideology; thus, safe spaces that allow for dialogue needs to be created for Yotsuba to sustain itself.

Sanbu network had institutional relations with consumers and private logistics companies, but private business ethics predominantly governed the relations. That is, they had an understanding with producers of input only as far as buying from them while they relied on logistical companies to ferry their products. Nothing went beyond that. There were no mutual understandings with other producer cooperatives, but instead, there existed some competition. The cooperative was also able to access information from various government sources which it was able to repackage to be palatable to the farmers. Those farmers that can access benefits from institutions outside their cooperative did so from their initiatives and less on the cooperative initiative. There are other cases where the farmer researches on their own and establish relations by themselves. The latter is most common. Many of the farmers highlighted that besides selling the product, the network was not helping the farmers in other ways. Their role was confined to the selling of outputs and provision of an organised outlet for small farmers.

This network operates almost like a private company; hence, it works with other private companies just like any way that other companies work with other companies in this area. It is all business-oriented. There are other networks almost as 'privately owned' as this one. Most are more prominent than this one and are only concerned with the marketing of produce. (Takahashi interview, December 2017)

Nevertheless, one of the younger farmers highlighted that they had reduced the amount of produce that they sold through the cooperative and suggested the need to re-evaluate intra-cooperative relations since the cooperative leader's committee or executive no longer reflected the interests of the membership. They lamented that the network was too concerned about making a profit (just like any other company) such that it was affecting the production itself from the farmers. Kobayashi Jnr. also seconded this:

> The network cannot survive for a long time without the support and effort of the farmers; thus, we need to make sure that the farmers are also happy in the arrangements between consumers, companies and the government. (Kobayashi Jnr. interview, December 2018)

The presence of the government in the Sanbu network became less pronounced after moving out of the regional JA cooperative. They feel like they do not need to conform to the Cooperative Act but rather to the Companies Act. Although this reflected some level of misunderstanding of Societies Act, it also indicates the value placed on the relations with the government. The formation of the cooperative model for Zimbabwe will consider this result.

In the case of JA Green Ohmi, considering that the government historically institutionalised the movement, it is complicated to understand where the mandate of the cooperative institution ends and where that of the government starts. However, the Abe administration has recently intensified efforts to change this as it wants to privatise the JA. The JA is expected to operate more as a company and reduce its 'dependence' on the state. These running battles between the state and the JA have given birth to self-reform policies which local and national cooperatives are supposed to carry out (see full discussion on page 193). The MAFF and JA-GO are jointly carrying out farm reorganisation[17] program in Ryuo. In some

---

[17] Farm reorganization is an on-going process that started in the 1950s where disaggregated pieces of land that belong to one family were reorganized through swapping, buying and selling scattered plots to make them more manageable and easier to cultivate with small tractors on. The current reorganization tries to

cases, retiring farmers approach the cooperative directly to surrender land, and JA-GO liaises with the MAFF during the processes.

> Some farmers give their land to the JA […]. JA's way of land rental is performing much better than the MAFF way of lending in this region […]. It is essential to keep contact and relationship with the government despite several disagreements. However, it is supposed to be neutral. JA has a political wing or association which is different from the JA production group. This wing has more contact with government politics and political institutions. (Kawakita interview, June 2019)

Interestingly, the JA-GO seemed to be hostile to private businesses, especially those that seek to be in agricultural production. JA-GO is part of the JA and hence has direct access to JA cooperative banking resources. This enables the cooperative to operate without too much reliance on private capital. Their rationale for categorically avoiding production partnerships is that agricultural production should never be privatised, and hence they refuse to privatise their local cooperative. The message was clear.

> There is no room for private companies, especially in rice production. There is no partnership in production whatsoever. Even in the future, there is no possibility. Unless the businesspeople come from within the membership. Rice production is a public venture and uses public water and hence cannot be privatised. (Kawakita interview, June 2019)

We need cooperatives to better negotiate with other sectors in the agricultural value chains. This sub-section revealed how necessary it is to keep all relations warm despite philosophical orientation or production approach. Thus, all cooperatives valued relations with the state, the private sector, and other cooperatives. In formulating my model for Zimbabwe, this will be crucial.

---

consolidate abandoned and under-utilised farmland (Hirasawa, 2014, p. 16; MAFF, 2017, pp. 7–10, 2018)

Management issues in the cooperative movement

The discussion thus far has highlighted several management issues and how they are supposed to be solved based on the data I collected in Japan. This section extends the analysis through the lens of the three cooperatives studied. The three utilises different types of management approaches, and all have varying effects on cooperative development. Management approaches refer to the method of leadership by the management committee and can fall into three broad categories. The first one is the autocratic approach in which decisions are carried unitarily without consideration of the lower echelons of the cooperative; it is swift. The *laissez-faire* represents the second category whereby the manager takes the role of a mentor rather than an authoritative figure and allows for lower echelons to take part in decision making. Decisions take more time to make in this manner. The third category entails other variations, which can be consultative, persuasive, or democratic means (Cornerstone, 2019). Ideally, because a group of people owns a cooperative, democratic, and *laissez-faire* type of management styles are expected to dominate the cooperative movement.

Based on the data collected from the Yotsuba cooperative, I found that the farm and the cooperative management styles were more inclined towards the *laissez-faire* axis. That was mainly because of the intense focus on cooperative ideology. Yotsuba itself was formed out of an earlier version of Kansai Purchasing Cooperative (KPC) because the older refused to change their ideology. In 1984, through a somewhat autocratic management approach, KPC called for an abrupt change from inorganic to pure organic ways of production mainly for the benefit of the consumers (Tohira interview, March 2017). However, consultations with the producers themselves revealed that they wanted a gradual change to allow their farm infrastructure to adjust into organic methods. This then led to the separation of the KPC into two; Kansai co-purchasing cooperative which exist until this very day but has primarily remained as a consumer cooperative, and Yotsuba.

Yotsuba focused on the producers and has been able to negotiate higher prices for the farmer's products illustrating the differences in management styles. The leaders of the cooperative added that:

182

A school has been established, which involves monthly lectures about the ideology and the history of the Yotsuba to try to teach the younger members about this way of life, and how we can help and sustain producers. Everyone is equal on the farm, they get the same salary, and the aim is encouraging equality and a more advanced level of living. It sets the farm miles apart from all other farms. This is only for Nose farm, other farms in the same cooperative can choose what and how they want to run their farms [...] people here (Yotsuba) treat each other well, and they appreciate the farm even though the money is little. The system seems inefficient since we always discuss everything before deciding. (Tohira interview, March 2017)

Although we buy expensive machines, such as meat processing machines, we need highly specialised skills and knowledge to operate them [...]. Our way of management is inefficient in terms of economic production because it takes time to make decisions, and not competitive in the capitalistic production system. However, we chose this way to bring up the next generation. Let us avoid an obsession in which everybody seeks to expand their business. What we should seek is how we can make a strong alliance among smaller local units of production, distribution and consumers. (Tsuda interview, March 2017)

The management styles of Sanbu network stands diametrical to that of Yotsuba cooperative. Based on interview data, it represents a consultative management system. The network has five executive directors and a standing director (representing all these directors). These directors were not democratically elected (Nakamura interview, December 2018). The farmers I talked to seemed to think that this was normal since the network did not belong to JA cooperative but was rather more of a private network. This issue seemed to be at the heart of the cooperative problems (see page 193). The result of such management system is that the management committee becomes unforthcoming in sharing a specific type of information or even listening to the concerns of the members. Nakamura had a lot to say on this issue:

183

The members of the cooperative should request to have a more open flow of information, especially information to do with the activities of the top leadership. Free elections of the executive should be democratic and well-constituted, and new people or new blood should form a significant proportion of the executive [...] Recently, they just decided to reduce the number of meetings (citing the need) to improve the quality of the meetings. To solve farming, there is a need to improve the decision-making framework and the flow of information between (from) the farmers and (to) the executive. The acceptance of such reform, of new ideas and of bringing in new younger farmers will save the network. (Nakamura interview, December 2017)

The problem in this respect is that many meetings that are held are pre-decided or pre-concluded. The actual meetings serve only to rubber-stamp the executives' decisions or as a way of informing the members on the decisions that have been reached by the management (Takahashi interview, December 2018). This is peculiar because Sanbu is not as big as JA Green Ohmi, which requires that management formulate solutions, and the members vote on their feasibility, Sanbu management is expected to engage more with the farmers. In the end, the members feel as if the network is a private logistics company organised as a cooperative to enjoy the benefits of cooperative law. Some farmers felt that since the network does not require any membership fees, and that farmers would only pay a certain percentage from the sales made through the cooperative. This means that the cooperative was not for the farmers but was for the leaders.

Additionally, some farmers argued that management style was conducive for what they called 'natural/core farmers' and not for new farmers. Natural farmers mean those who inherited a farm or have always been into farming since they were born. New farmers were those that once worked in other economic sectors and had decided to pursue agriculture. Thus, natural farmers did not care about calculating the actual net income that they were getting since they had lower cost as opposed to new farmers that have to pay for land rates. In recent times, the Japanese agrarian structure has seen an increase in the number of these non-natural farmers. To fix this emerging problem, cooperative leadership needs a representative of the new farmers during decision making.

The management type of JA cooperatives is typically a top-down approach. The latter has a lot to do with the Japanese system of vertically structured society popularly known as the *tate shakai* (縦社会) than it has to do with the cooperative as discussed in Nakane (1967, pp. 58–64, 1970, pp. 40–48). In a vertically structured society, seniority plays a significant role in decision making, and the opinion leaders can manipulate the strength of the voices of a group of people. However, as we saw in Yotsuba, this can be avoided. Also, to a greater extent, I found out that the JA-GO had reduced the intensity of this top-down approach, probably from an autocratic management system to a more persuasive one (Ishida interview, July 2017). This finding is buttressed by the fact that there still were many productive activities (agricultural production, marketing, inputs acquisition and output marketing) that were observed at the JA-GO facilities than in any other JA place that I had visited. The JA facility in Nose and that of Sanbu area looked like ghost shops in which no farmers ever come to purchase inputs or to seek advice from the cooperative structures.

Although the management of the JA-GO cooperative is the typical JA style of cooperatives, it has been able to encourage participation by the farmers. The institutionalisation of the movement in that area was unavoidable because JA has strong governmental ties. These ties were forged during the development phases of the cooperative in which the state needed support from the rural population while the people wanted protection. One way or the other, it would have happened, and at some point, it is desirable, especially in the initial stages of the development of the cooperative movement. These lukewarm relations have severed in current times which has seen regional cooperatives such as JA-GO re-connecting with the farmers for support. JA-GO is a multi-purpose cooperative just as in Yotsuba, and unlike Sanbu network which is a single-purpose cooperative. Single-purpose cooperatives are most suited during the initial stages when the cooperative is small.

The cooperative can open up and become a multi-purpose cooperative when it grows bigger. If a cooperative has many business activities in case the organisation faces a downturn in one section, it can use other businesses to stay afloat. As in the case of JA-GO, it has been involved in the insurance and banking sector to fund some of the non-profitable activities like crop input supply and output marketing. It is essential to understand that cooperatives, especially in the case of Japan, used profits from banking and insurance to carry out cooperative farming activities. While this approach worked out for Japanese movement, developing countries (with limited access to external financial capital) can begin by SACCO type of self-help financing institutions with cooperative banking as a long-term plan. However, they should avoid getting carried away and forgetting to use that money to develop their members. To minimise this, JA-GO always has meetings with the farmers and the consumer organisation. During data collection, I managed to sit in one of the meetings to witness the negotiations for a price increase in the price of vegetables sold to a consumer cooperative.

The political economy of cooperatives in Japan

The role of the government in the cooperative movement in Japan is well documented. As discussed in Chapter Three, the state was concerned with securing a stable supply of food, development of the agricultural, forestry and fisheries industries. Given that cooperative socio-organisational structures dominated the rural areas of Japan, it was probable for most government programs to follow cooperative organisational structures. It is vital to understand how different governmental socio-economic and political pressure affected the cooperative movement. Pre-war and post WWII cooperatives took part in political campaigns and even ran for office. In doing this, they increased many points of conflict with the ruling political echelons. Thus, after the war, all labour and farmer's unions were discouraged ahead of the more reasonable and easily negotiable cooperatives.

Yotsuba showed that it was not only difficult to separate politics and cooperatives, but that it was a mistake to separate them. This is the reason why they still had a politically active wing within their

186

cooperative structures and why they maintained a 'loose political wing' for activist and anti-government policy demonstrations. They had many problems with state policies, and there was a relatively long distance between the cooperative and the government or the private sector. In principle, Yotsuba stands for the exact opposite of the state's model of economic production, although an uneasy coexistence is evident between the two. All of the founders of the cooperative were influenced by the Cultural leftist ideology of the Chinese Revolution. In the past 40-50 years, Japan enjoyed economic growth, and the younger generation did not think much about underlying social problems. However, capitalistic economy has been in a perilous situation over the past decade or so. This has reignited their will, and they are more convinced that collective action to resist further economic liberalisation will be the saviour of Japanese society (Tsuda interview, March 2017). However, the interviewees highlighted that Japan has one of the most conservative societies who are opposed to radical changes that go against the central governments.

The 'loose political party' seeks to educate people and spread their ideas of an ideal way of life. The long-run objective is to try to set an example starting from Nose farm. People are not forced; they have to observe the Nose's way of living, and hopefully, they can change their perception and their way of thinking. It obviously will take time, but the Yotsuba leaders highlighted that they are determined to continue working in this way (Tohira interview, March 2017). Several farmers highlighted that they are disgruntled by the way JA cooperative is handling agricultural development in rural areas. However, they still retain their membership because they are afraid of radical change by withdrawing their membership from JA abruptly. Another farmer explained that the problem was not really with the government but in the structure and organisation of Japanese hierarchical rural society which made it easy to manipulate authority. In addition to this, the farmers also hold an image of the goods times that the JA and the government had brought to them during the Japanese 'golden ages' and hence, the older generation sincerely held high hopes of reliving those days (Tohira interview, March 2017). They bewailed:

In this local (rural) region, we are still strangers after these forty years, as some residents regard us as a radical (dangerous) leftist organisation. We sent members of our farm to the town assembly for five terms, that is, twenty years. One of our members stood as a candidate for the town mayor, but we were defeated because we are still outsiders. (Tsuda interview, March 2017)

Outside of the farm, the cooperative organised an anti-nuclear demonstration in Osaka for the government to stop the production of nuclear electricity. They argued that there was no need to produce nuclear electricity because Japan could effectively meet its demands using other safer methods. Their counterargument is that nuclear energy investment was to make supernormal-profits at the expense of the safety of the people and the environment. They cited the Fukushima nuclear accident repeatedly in their argument. I watched some documentaries about the disaster during my stay at Nose farm. The documentaries were bought to help members understand the consequences of chasing profits ahead of human safety. The political wing was also involved in another anti-governmental protest which it thought would not benefit the environment or the rural people as reported below:

[…] I am actively involved in the political wing of the cooperative, tomorrow I am going to demonstrate against conspiracy law under discussion in the Diet. Am going to demonstrate at the Nose supermarket and also later in the Osaka railway loop line [and eventually] to demonstrate at every station. (Tohira interview, March 2017)

The law passed through the parliament three months after my interview with Tohira amid remonstrations from within Japan and multi-national organisations (like the United Nations). Such a cooperative requires more time to observe the dynamics than was possible during this research because of resource limitations. There are always problems between different echelons in any organisation. However, Yotsuba had a seamless integration of decision making and development strategies within the leadership, the management and the members/workers. What this meant is that inter-echelon contradictions were, at the best-case scenario, minimised or in worst case scenario deeply hidden. Higher levels of information sharing, openness and humanness seemed to iron out political as well as social discontent within the movement.

On the contrary, Sanbu network members agreed that it was tough to separate the cooperative movement from politics since most of the members were integrated into the same through voting in local and national elections. However, they seemed to be in line with most of the government policies and programs and did not mind an even more pro-active role from the government. They had a considerable amount of trust in the government's thrust of economic production, including mild support of the pro-liberalisation approach to agriculture. The network members highlighted that the government had several programs that it ran through the cooperative to benefit the farmers and this was, in fact, one of the reasons why some new farmers were attracted to the cooperative (Yamada interview, December 2018). Some members of the network had this to say:

> The politics and economy are united. The economy cannot solve politics sometimes, but politics can solve the economy all the time. Thus, the cooperative cannot be divorced from politics. Politics can activate and change the farming situation. (Kobayashi Jnr. Interview, December 2018).

> There is no way to stay out of politics because it affects everything else. (Nakamura interview, December 2018)

As highlighted throughout this book, the level of state involvement is instated by the active JA national movement involvement despite deteriorating relations. For the state, Green Ohmi type of regional cooperative continues to be appealing to the government because it can still push development and farmer assistance programs via this same channel. Maintenance of a friendly JA cooperative structure also enables the government to keep such organisations as Yotsuba in check. This is so because if the government completely abandons the cooperative movement, it surely will give the farmers freedom to form genuine bottom-up organisations that have the power to question ruling party decisions as seen in the Ugandan case (see Wedig & Wiegratz, 2018).

Developing countries like Zimbabwe needs to have government structures that work (and sometimes overlap) with those of the cooperatives to cascade development initiatives to the cooperative members collectively. I asked the interviewees on their opinion of who had ultimate power in the JA movement, the government, or the cooperative members themselves. They highlighted that on the surface, it seems like the government has overwhelming power. Still, there is a need for a deeper understanding of inter-JA politics to answer that question in its entirety. An academic who has studied cooperatives thus opined:

> Because the government has institutionalised the cooperative movement, it is hard to understand which side, the government, or the JA, possessed real power. However, through working together, it may be safe to say that they worked together on a 50-50 basis to support the farmers. (Ishida interview, July 2017)

In conclusion, while observing the difficulties in being apolitical, and that it may not be desirable to be, it is crucial to avoid politics within the initial stages of the cooperatives when it does not have either the intellectual or financial resources to be politically active. Instead, the cooperative should strengthen its institutional structure such that when political players and private business approach them, they are better prepared to negotiate or bargain deals that benefit them as well.

Nature of amendments to the Japanese Agricultural cooperative law

As discussed in Chapter Three, the Japanese government has been trying to privatise agriculture as it did to other economic sectors such as the Japan Railway (JR), Nippon Telegraph and Telephone (NTT) and Japan Postal Bank (JP). While experts hailed privatisation in these service industries as a miracle success, critics remain adamant that privatisation in agriculture would not work (see more in Chayanov 1991). In the case of Japan, where the farmers control agriculture through their membership to the JA, privatising 'already private' cooperatives was always going to be problematic. Because the members have consolidated voices which they can air through the cooperative, they have resisted privatisation (degeneration) of agriculture for over forty years.

Latest amendments to the Agricultural Societies Act

The role of the cooperatives is still vital for rural development in Japan. The farm restructuring programmes heavily rely on negotiations with cooperative leaders and also seeking their help. As such, the 2015 revisions, while trying to create more market-oriented cooperatives, which operate within the laws of the free markets, the government still recognises the strength cooperatives in redistributing profits to its members (MAFF, 2018). The most positive part of the amendments is the stipulation that only certified farmers or those with proven agricultural practical skills can occupy positions in the management committee and board of directors. This is very important in order to improve management and leadership qualities, areas in which cooperatives have often been criticised to lack.

However, there are several parts of the intended amendments that force the cooperative to degenerate. Cooperatives can no longer force members to use cooperative channels for marketing their produce. Also banned in the new law is the forced use of farmer's dividends as cooperative capital. This means cooperatives must pay out dividends to the members, and then members can then decide if they want to re-invest them or not (Nishikawa, 2015). This is a movement away from strict Raiffeisen principles.

Additionally, two or more cooperatives that specialize in the same product can now be formed to give more freedoms or options to the farmers to do what they desire. Leaders of the cooperative should be the most qualified in the membership and should be a certified farmer. Cooperatives will now freely reorganize, dissolve or form new ones and convert into joint-stock companies. The change of the JA-Zenchu into an incorporated association was instituted while prefectural associations and JA Zennoh were given the power to decide on whether they would want to remain as cooperative bodies or convert into corporate ones. Larger income making cooperatives (business income sources) are now supposed to get audited by reputable or government approved/recognized auditing companies and not just by the JA-Zennoh and the JA-Zenchu. I used these lessons to construct a new cooperative framework for Zimbabwe in Chapter Eight.

Although the Agricultural Cooperative Law interfaces with other laws inside the MAFF and other state institutions such as the Food Agency, the treatment of cooperatives under the Anti-Monopoly Act of 1947 is of great importance. This Anti-Monopoly Act was put in place to prevent private monopolization and excessive concentration of resources into a few hands. Japanese cooperatives enjoy special treatment concerning the anti-monopoly law. The law seeks to restrict, as much as possible, the prevalence of unfair and unreasonable restraints on production, sales and transaction of production inputs/outputs (GoJ, 1947a, p. 1). Farmers and agricultural cooperatives are described or recognized as enterprises within this law. Article 22 of the antimonopoly highlight cases in which cooperative enterprises (mainly small-scale enterprises) are exempted from the restrictions imposed on other businesses.

However, a cooperative may not be protected by this law if they *i*) forbid farmers from using other channels (including private channels) to sell off their output or to source their inputs *ii*) force members to only use cooperative facilities only if they marketed their products through cooperative channels and, *iii*) grant loans and credit to members on the condition that these farmers sell their produce or outsource their inputs through the cooperative channel. In such cases, the respective unit cooperative is not protected. The same situation holds at the national level. No preferential treatment should be given to the unit cooperatives that have more business activities with the JA-Zennoh (Godo, 2015; GoJ, 1947a, p. 17). I note that for cooperatives in the developing country like Zimbabwe, adhering to such a law system would undermine cooperative development.

192

Farmer views of cooperative reform

The pressure to reform the cooperative has intensified in recent decades. Because of the strength of farmer agency exerted through cooperative structures, they negotiated reform on their terms; this is what has come to be known as JA's so-called 'self-reform'. It is vital to analyse how the farmers in JA-GO has been carrying out such a reform and the socio-economic sources of agency. JA-GO represents the general opinion of the cooperative leaders towards reform, not completely hostile to it, but nevertheless unsure if the objective of the reform is intended to build the cooperative or destroy it.

For JA-GO leaders and members, the reform has two fronts, at the national association level (structure, ownership, and legal governance) and also at the district/local level (improved production and management levels). The JA-GO members felt that the actual reform as advocated by the government was targeting a reformation of the national cooperative associations. Due to JA national association solidarity, JA-GO has the prerogative to decide on which parts of their cooperative to reform and which ones not to. Through a consultative process with their members (meetings, seminars and representative discussions), they made provisions, proposals and policies to *i)* improve social and agricultural development *ii)* adopt business models in marketing agricultural products and *iii)* improve cooperative human resources including fostering the next generation of farmers:

> To be honest, at the local level, reform is an iterative and gradual process that we have always done even before the 'reform'. Issues are discussed between members and their respective JA management committees. It is taken further to the meeting of representatives, and they will take the issue to the annual conference. Because JA is expansive, a plan [for self-reform in the case] is prepared by the top management in consultation with the representatives, then discussed and improved at the conference. (Interview with JA-GO leaders, July 2019)

Many JAs in Japan are engaged in wide-ranging business (banking, insurance) which compete with private businesses. However, significant portions of their income still come from agriculture-related activities; some revenue comes from membership and extension services fees. Although the cooperative's portfolio is rather vast, business activities (banking and insurance) provide most of the income, which is in turn used to finance other not-so-profitable activities such as extension and community development. This is listed as a way of beneficiation for the members and represents self-reformation. Beneficiation within JA-GO also includes annual dividends, and other non-direct means such as input price reduction (subsidy). One per cent of the membership fees is also given back as dividends.

Of significant importance was to understand local opinions about changing the cooperative structure to become an investor-owned company. My interview data showed that there was no place for private companies in production and owning agricultural land. JA-GO would only work with private companies in marketing transactions with producers of inputs and buyers of products, that is, the producers of inputs (fertiliser and equipment) and buyers of produce (consumers, supermarkets). The recent push by some of the big supermarkets to produce on their own was an attack on the small-scale farmer's livelihoods. The leaders of JA-GO vehemently highlighted that:

> JA-GO has no plans to become a farming company; there is no room for changing into a private company, or even partnering with a private company, especially in the production of rice. Even in the future, there is no possibility. Unless the businesspeople are the farmers themselves producing within the cooperative. Rice production is a public venture and uses public water and hence cannot be privatised. The target of the reform is JA-Zennoh and not necessarily the regional cooperatives that are involved in the production. We want to serve the farmers indefinitely, and we will continue self-reform to keep production and incomes up for the farmers as we have always done. (Kawakita interview, July 2019)

Continued domination of the cooperative, which supports less than 3% of the population has been argued to infringe on trading rules and regulations that call for total deregulation in agricultural markets. Thus, from JA-GO perspective, the government-advocated-reform seeks to liberalise the agricultural markets to protect the IT manufacturing sector. However, due to the strength and power of the national movement, they realised that they need to navigate the issue with extreme care. Kawakita pointed out that:

> [...] the government-led reform benefits the industry and the overall economy, but the JA at the local level is focusing on benefiting the farmers. That is the most remarkable difference between the two reforms. The good thing is that farmers were allowed to decide their reform agenda [...]. JA-GO has a political wing or association which is different from the JA production group. It tries to have certain politicians into the office of national Diet and prefecture. (Kawakita interview, July 2019).

As argued in Chapter Two, if the national cooperative association is allowed to turn into a company, the cooperative will start to degenerate. The passing of a law to pressure JA-Zenchu into a company is a direct attack on the cooperative movement. JA-Zennoh got the option to decide whether they would change into a company and they refused. However, based on the global neo-liberal thrust, it would be tenable to say the next amendments to the cooperative law will make it mandatory for JA-Zennoh to degenerate while granting local JAs the prerogative. This might go on until it becomes mandatory for all cooperatives at the local level to become companies. The end game in all this is for the private businesses to enter into the lucrative business of agricultural production. However, the evidence shows that the JA has the potential to resist at the local and national level. In an opinion piece circulated on the JA communications website, Mr Morishima lamented:

> On the surface, it seems that the simple objective of reform is to link the business community with all agriculture. However, it is not the actual purpose. The real strategic purpose is for capital to take over the agricultural cooperatives. [Besides,] it does not take over the entire organisation but only for profitable businesses, then cut off and discard

those that are not [like extension]. That will dismantle and collapse agriculture.– (Morishima, 2017)

The government justification for a top-down approach to reformation usually cites the changes in membership structure for JA Green Ohmi. Scholars have widely accepted (and some cooperative members as well) that the cooperative focuses more on the needs of the associate members at the expense of the core/regular farmers (Key informant interviews, 2017-2019). Over ten years from 2008 to 2018, associate members increased from 11,806 to 15,222 while full members reduced from 9,087 to 8,681 bringing the proportion of associates to 64% of the membership (JA-GO, 2019). Resolving this contradiction for the cooperative has proved to be difficult thus far because the associate member business activities have driven the profits for JA-GO which it then uses to fund the rest of the less profitable portfolios like farm-guidance – extension (Ishida interview, July 2017). It is critical to point out here that the rate of ageing/retiring farmers influences the reduction in the number of full members and not dissatisfaction in the cooperative *per se*.

Active participation and involvement of all members in cooperative affairs is one of the critical things that the movement should safeguard throughout the reform. Interview data revealed JA-GO was still focusing on the regular members and that the regular members, who had voting rights, directed the process of reform. Thus, associate members had no voting rights although their concerns are heard (because they also contribute immensely to incomes) and considered during program formulation. A reduction of full-members and a stable employment rate has meant a decline in the member-staff ratio, hence shortening the distance between the JA-GO workers and the farmers. Altogether, this had improved service delivery to the membership, JA-GO leaders highlighted that the bulk of the workers are members of the cooperative and that even some specialist skills such as auditing were done within JA at the national level to reduce external exploitation of labour. Self-exploitation is critical to reducing cooperative degeneration.

Agricultural cooperatives control a lot of finance capital in Japanese agriculture. High potential returns to investments probably explain the push by external players outside Japan who want to enter into the market that is currently controlled by the cooperatives. They also want to tap into the cooperative capital base.

Contemporary issues in Japanese Agriculture

There exists a myriad of challenges in the contemporary cooperative movement of Japan. The intensity of challenges affected the organisational structures of the three cooperatives under study in varying ways. These challenges, which included an ageing agrarian structure, scattered/disarticulated farmlands, and cooperative management issues, are elaborated below.

The national problems: Ageing agrarian structure and disaggregated lands.

The general trend in the rural population continued to show a decline between 2010 and 2015, which translated to decreases in the number of people employed in agriculture. The rate of change in agricultural employment was positively correlated to that of change in rice production, suggesting that the majority of the labour was employed or were involved in rice production (Figure 5.4).

Figure 5.4: Employment trends in Japanese agriculture (%, 1970-2014)

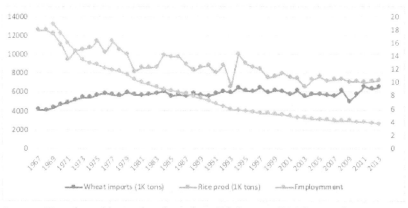

Source: Developed by author based on FAO stats (2016) metadata

On the other hand, change in agricultural labour is inversely correlated to the amount of imported wheat; from 1988, a decrease in the total amount of agricultural labour is offset by an increase in the amount of wheat imported into Japan. Although the high demand for labour in the industry (leading to high rural-urban migration) explains the decrease in the employment rate in the sector, the decrease reveals a critical agrarian issue. Not only is the number of producers decreasing, but the majority of the remaining farmers are old aged. The number of farmers aged 70 years or older accounts for about 50% of the total agricultural workforce (Japan Times, 2016). Thus, one of the most significant issues facing JA is how to attract younger farmers. The future of agriculture lies with the younger farmers as they are the ones with the energy and are more likely to assimilate technological advancements than older farmers.

Japan's rice productivity was approximately 40% lower than that of California in 2012. This was because the government, through the Minister of Finance and the JA, indirectly instructed MAFF not to develop higher-yielding rice varieties that may disturb the price stability and may reduce the price of rice and its associated sales. This is the political economy of rice and agriculture in Japan which is meant to keep the prices of rice higher and protect the farmers (George-Mulgan, 2005; Yamashita, 2015a, p. 72). Given the difficulty involved in bequeathing a piece of land, and the rate at which older farmers are retiring (or dying) and the number of non-farming households, the proportion of abandoned farmland has soared to 50% of all agricultural land in Japan (Nishikawa, 2014, p. 1). The government faces a dilemma as to whether to the current status quo can reform itself or allow accumulation of land by those that can utilise it (and risk a potential 'reversing of the 1945 land reform'). Land needs to be amalgamated into larger parcels, which often attracts commercial entities including foreign multi-national agribusiness companies.

Local-level constraints in the cooperative movement

There exist some generic production challenges such as the high cost of inputs, unavailability of stock feeds in Japan, the effect of an ageing farmer's population within the cooperative movement. For

Yotsuba type of cooperatives, a number of the older farmers relied on the cooperative labour arrangement systems to continue agricultural production. Some farmers offered their labour to help older farmers harvest their plots, and this labour was arranged and organised through Nose farm:

> I think our position is precarious. We cannot continue with our way of production, and people are getting old, and also the environment and the economy is changing, so there is need for more effective methods to attract more young people. We need to build relationships with the younger ones. (Tohira interview, March 2017)

These issues give rise to other issues. It takes more time to convince the younger generation to adhere to ideology and for them to understand its importance fully. One of the issues that they faced in terms of following a clear ideology is the fact that their cooperative was growing bigger. As the cooperative grows large, the more difficult for people to adhere to the ideology (Tsuda interview, March 2017). At the same time, an ageing population forces the cooperative to rely more on machines and hence reduce the number of people working at the farm at any point in time. On the other hand, there is also a growing proportion within the older farmers. They wish to abandon organic farming citing the fact that it was labour intensive and also that the premium for organic vegetables had drastically reduced over the years, reducing the incentive to grow them.

In **Sanbu Yasai network**, many challenges existed, ranging from intra-cooperative contradictions to those that were more external. Generic problems that affect the whole agricultural system in Japan, such as ageing, was listed as one of the most significant challenges in the cooperative.

> Generally, the movement is becoming less active, which, to a less extent, can be blamed on the overwhelming proportion of ageing farmers. The older farmer's do not have the capacity to innovate. It is the biggest challenge in Japan now. I am worried, even more than the network, about the successors of the current older farmers. If young

199

farmers increase, the network will be rejuvenated naturally. (Kobayashi Interview, 12/2018)

I think that 50% of the executive should resign and be replaced by younger farmers who understand the plight of non-natural farmers or younger farmers. New farmers that are joining do not know about the other nasty stuff. Also, the office people are not farmers, and they do not know the plight of the farmers, and they do not know the realities of the farmers. They are a bit detached from the realities on the ground. (Yamada interview, December 2018)

However, ageing farmers comes as an opportunity for some ambitious young farmers like Nakamura, who was interviewed during the research. Nakamura represents one of those younger farmers who are desperate to change the cooperative *modus operandi* because he firmly believes that the movement has a higher potential. Still, it needs leadership renewal, and it needs to adapt more to the changing environment. His story is presented in *Textbox 5.1*.

One of the most significant observations from Nakamura's story (in *Textbox 5.1*) is that a few individuals in the movement now think and believe that they own the network on the premise that they formed the network in the first place. These insinuate that they have a right to enjoy extra benefits from the cooperative and that the movement belongs to them. Eventually, the other farmers/members lose faith in the cooperative and begin to leave the organisation.

According to the interviewees, the organisation was starting to lose sight of their objectives because of their need to make profits coupled with strict management styles that do not allow for flexible decisions to be made without the approval of the top leadership (Watanabe interview, December 2018). This represents Sanbu Yasai cooperative's foremost challenge.

Other farmers, especially the younger ones, seemed to relate to Nakamura's story somewhat inadvertently. They highlighted that the flow of information was unidirectional, and they had no significant input in the decision-making process. Information flow is a symptom of the more significant challenges that dominate Nakamura's story. For them, the reduced flow of information was the tributary source to the river of other challenges such as the reduced number of consumers willing to buy their products. The other farmers say:

> I am not satisfied with the flow of information now. In the beginning, we had managed to strike a deal with the consumers, and the deal was based on the smooth flow of information, which is no longer the case. We had a closer relationship with the consumers then. (Kobayashi interview, December 2018)

> The flow of information is OK, but the quality of information may not be. While information such as sales, losses and prices of produce is shared widely, such information about the rationale of some of the decisions made by the executives is not shared and is kept within the top leadership. (Takahashi interview, December 2018)

Overall, challenges will always be present in the cooperative movement, and this sub-section shows how important generational renewal is at the cooperative level to ensure continuity and to bring in new ideas and dynamism. The challenges can be resolved through a stronger emphasis on education and philosophy as in Yotsuba, or improvement of information flow and democratic means as in Sanbu. The challenges of JA-GO are discussed in the proceeding section.

Textbox 5.1: The story of Nakamura from Sanbu Yasai Network

As told by Nakamura san
I used to work for a company that dealt with transistors and semi-conductors with head offices in Tokyo, Germany, USA, and China. The job had an excellent salary, but it had much pressure which was not healthy for me. So, I quit and came to start farming. I got training from an uncle to start agricultural production 8 years ago. I managed to secure land from an old landowner who stays in Tokyo, who is too old to do farming. I had the option to [own and] pay for the land over 50 years, but I opted to rent it instead. It was way cheaper since I had to pay ¥150000 per hectare per year. I secured perfect [piece of] land, rich in black volcanic soils that is good for all types of crop production from carrots to daikon to peas. The sad part is that not many younger farmers want to come and do farming like me. A lot of the younger farmers are comfortable working day jobs in the city.

The network is facing several other challenges chief among them the lack of creativity and leadership renewal strategies. I believe that the era of organic food is slowly dying and that all organic-vegetable producing cooperatives should be prepared for the time when it comes to an abrupt halt. This is because consumer prices have been driven down such that people are not willing to pay the extra money for organically produced vegetables. Now I only sell about 55-60% of my produce through the cooperative and then sell the remainder through other channels such as by myself in Chiba supermarkets and restaurants. I thought about getting out of the cooperative a couple of times but stopped because of my respect for my mentors and his neighbours who belong to the network. A few farmers have stopped their membership from the network entirely because they do not see any profit to it.

However, the leaders or the members of the administration of the network seems to differ in thinking. They seem to think that organic food production is still viable and that the network should and will survive in the future. There is no advantage now when selling through the cooperatives since we are unable to sell at high prices. There used to be an advantage a long time ago. It does not want to change or to get creative people to change things within the cooperative, that is its greatest weakness. They don't make elections for leadership roles of the cooperative and hence, I have never been in the decision-making structures. I am only a member to help my friend who taught me about farming and to keep the connection and harmony within the village. The natural-born farmers have no problems with the way things work in the network because the government supports them, and they receive several subsidies. However, some farmers who are into farming as a business have issues since they want the cooperative to change into a more business model.

The disadvantage of the cooperative is contracted produce and the fact that the price gap between organic and the other products is very few. The other issue is if the public market price is low, then we cannot sell to these consumers because they would buy in the public market despite the contract with the cooperative. The leaders of the cooperative cannot force the consumers to buy.

Summary and conclusion

This chapter, together with Chapter Three, provides several lessons for the Zimbabwe cooperative movement, discussed further in Chapter Eight. In Yotsuba, they were more attracted to the ideological sides of the cooperative and less about making money. In Sanbu, although the concept of organic farming had played a role in members' willingness to join a cooperative, it was not enough to maintain their membership and the farmers were considering leaving the cooperative because it had become unprofitable. In Ryuo, farmers joined the cooperative as part of the national drive and had stayed in the cooperative because it had assured high incomes and access to various government programs, among other reasons.

In conclusion, while Yotsuba type of cooperatives, drenched in ideology, had been fashionable before the end of the Cold War, the overwhelming neo-liberal macro or global economy has presented several challenges to it. Adoption of these may be difficult in Zimbabwe. On the other hand, smaller business-oriented networks such as the Sanbu, which gained traction after the Cold-War (last 15 to 30 years), have also faced an array of challenges that threatened its sustainability. The JA-GO type of cooperatives also has many challenges. Still, they seem to have managed to navigate through the unfettered markets to become one of the most profitable enterprises in Japan. Their previous cosy relationship with the state and its political machinery has become adversarial cooperation. While partnering the private sector in input and output markets, the antagonistic collaboration solidifies farmer's agency in the rural economy. Such cooperatives could succeed if adopted, modified, and implemented in Zimbabwe. Thus, in my attempt to create a cooperative model for Zimbabwe, one of the questions I had to ask was what role ideology should be able to play in the cooperative frameworks. My study of the Japanese cooperatives reveal that it should be somewhere between Yotsuba and Ryu JA Ohmi cooperative

In the next chapter, I turn my attention to Zimbabwe. The main objective is to examine the nature and structure of the intra-cooperative environment of the Zimbabwe movement, as well as the external environment (government policies, laws, institutions and other relevant stakeholders) that influence its direction. The next chapter (and Chapter Seven) provides information that enables us to understand the state of cooperativism in Zimbabwe and the future opportunities for growth.

Chapter Six

Field evidence; Current Trends and Patterns in the Zimbabwe Cooperative Movement

Wisdom is like a baobab tree; no one individual can embrace it.
– African Proverb

Introduction

In his youth, grandfather had travelled extensively around several parts of Zimbabwe. He used to travel for work, sometimes for entertainment, other times to seek spiritual healing from African spiritual figures scattered across the country. One time he told us that he joined a boxing club as a boxer and travelled around the country winning bouts. With a crooked smile on his face, he told of the several young girls who fell for him after every win. Perhaps this is how he had met his first wife from Matabeleland. It is during these travels that he observed the countryside and how people lived, farmed, and reproduced. However, as he grew old, so did the changes in places that he had visited. He had not witnessed the changes that the FTLRP had brought to the countryside since 2000. It was then my turn to narrate to him my experiences as a young agrarian scholar. If he were alive, I would give him this book and tell him 'grandfather, all you need to know about the post-FTLRP is in here'. When we meet again, we will continue where we left off.

This chapter mainly utilised qualitative data collected through case study survey methods. By using qualitative data, we opened up spaces for comparison between Zimbabwe and the case study surveys from Japan. Therefore, just as I did for Sanbu, Yotsuba and JA Green Ohmi, I collected reactions and opinions of cooperative members (farmers) and leaders concerning the cooperative appeal, management, ideology, legislature, political economy, and the challenges.

The cooperative movement in Zimbabwe is viewed from several different perspectives. Some scholars view it as a powerful development tool (Romdhane & Moyo, 2002), and others believe that it is a spent force which has been tried, tested and has failed before (Akwabi-Ameyaw, 1997). Perspectives also differ by institutions; some Non-Governmental Organisations (NGOs) always utilise agricultural cooperatives to advance rural development, the government has continued to view cooperatives as an arm to control the peasants while the business community have viewed it as an antagonist. However, cooperatives should not be viewed as an opposing force to private business *per se*; instead, it should be viewed as a means for the peasantry to better negotiate with the brutal markets. The qualitative data used in this chapter was collected predominantly through key-informant interviews with experts and farmers as well as interviews with ministry officials. FGDs with cooperative leaders and farmers was a source of qualitative data as well. Content analysis was used as a qualitative analysis tool to study the Cooperative Societies Act (1996) and its proposed amendments.

Characteristics of survey participants in Zimbabwe

As indicated earlier (Chapters One), I managed nine interviews and two focus group discussions. Some of the names of government officials and other interviewees have been changed upon their request. Thus, in this section, I analyse the information provided by interviewees, as shown in Table 6.1. Focus Group Discussions have carried out with *i*) Chairpersons of agricultural cooperatives from Mashonaland East Province which consisted of 18 chairpersons, and *ii*) Managing committee of Tagarika Irrigation and Tractor Cooperative which consisted of 5 persons. The general idea for this chapter was to let the farmers speak by listening to their narratives as a way of amplifying their voices.

The respondents from the Zimbabwe case study were farmers drawn from 9 cooperatives in Goromonzi district. Seven of these cooperatives were in the CA and were involved in a variety of agricultural production activities ranging from organic gardening to egg production and dairy milk production. These were Chikwaka

Dairy Cooperative, Gosha Eggs Cooperative, Gutu Golden Eggs, Kumboedza Cooperative Gardens, Shungu Organic Cooperative, Simba Ivhu Cooperative and Survival Skills Cooperative. The other two cooperatives were in the resettled area (A1) with the predominant activity being field crops and horticultural production; namely Tagarika Irrigation Cooperative and Xanadu A. Agricultural Cooperative.

Table 6.1: List of structured interview participants in Zimbabwe

Name	Date and place of interview	Cooperative Name	Role or occupation
1. Cooperative registrar	March 2018, SMECD* HQ in Harare	SMECD*	Cooperative Registrar
2. AR Mada			Assistant Registrar (AR)
3. AR Nyamunda			Assistant Registrar (AR)
4. ER Mavhinyingwa	March 2018, Mavhinyingwa's house in Harare	Former SMECD	Retired Registrar (ER)
5. Mrs Maziva	February 2018, CACU HQ** in Msasa	CACU	CACU Chairperson
6. Dr Matondi	April 2018, Skype interview	Ruzivo Trust (Research Organisation)	Scholar
7. CP Manthosi	March 2018, GHHCU*** Offices in Harare	GHHCU	Cooperators
8. F Mugaba	February 2019, Farmers' homestead in Xanadu	Xanadu	Farmer
9. F Manyika		Xanadu	Farmer

* = Ministry of Small & Medium Enterprises and Community Development (SMECD); ** = Head Quarters; *** = Greater Harare Housing Cooperatives Union

The most used words within interviews and FGDs differed depending on the role of the interviewee within the national cooperative structures. I divided interviewees depending on whether they were farmers (members and cooperators) or experts (scholars and ministerial officials). The experts mainly spoke about the cooperative as a government-controlled movement within the concept of the national program (the word 'Zimbabwe' in *Figure 6.1*). They also used other buzz words such as 'model', 'development' and 'ministry' more frequently than farmers.

Differing perspectives emerged from the farmers who talked a lot about community orientation (the words 'farmers', 'people' and 'members' in *Figure 6.1*). Additionally, they talked about the 'movement', 'equipment', 'meetings', 'tractors' and 'the government'. These differences are noteworthy because they give us an idea of the context and nature of the challenges and solutions they face in the cooperative movement.

Figure 6.1: The most frequently used words by Zimbabwe farmers

Source: Compiled by the author based on own survey data, 2018-19 (NVivo word cloud query)

Overall, in conjunction with 'government', farmers talked about 'movement' and 'the people'. The Zimbabwe cooperative movement is, therefore, much affected by the interaction between the government and the farmers. In this sense, it was vital to try and understand the political economy of agricultural cooperatives in Zimbabwe.

Overall, across all the nine interviews and 2 FGDs, the most discussed themes were 'challenges' followed by 'solutions' and the 'political economy' of cooperatives in Goromonzi. Other issues that constituted the discussions were issues to do with the 'management' of the cooperative, issues to do with the 'independence and sustainability' of the movement, 'ideology', 'appeal' and issues that affected the formation of the cooperatives (Figure 6.2). Noteworthy from the qualitative data results was that although many challenges persisting in the cooperative movement were highlighted, the farmers and the experts knew or had some idea about solving them. However, they could not act on their plans because of unconducive management and governance environment. Some issues (such as corruption and abuse of cooperative facilities) could not be solved because they had a lot to do with the political economy of the cooperatives. The sections that follow discuss the eight themes in greater detail.

Figure 6.2: Number of times an issue was mentioned in interviews and FGDs

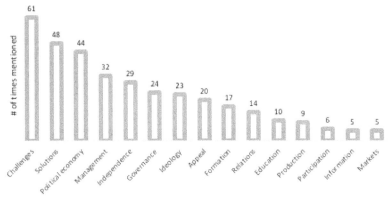

Source: Compiled by the author based on own survey data, 2018-19 (NVivo matrix coding)

Using NVivo's cluster analysis, I statistically examined the relationship between the chosen themes using Pearson's correlation coefficient method. Each time a respondent spoke about governance issues (e.g., the legislature, rules, and laws), they also talked about the independence of the cooperative movement from the government. This fact means that the source of threats for cooperative independence stems from the legislative framework or the Cooperative Societies Act of 1996. At the same time, issues of management of cooperatives were discussed within the concepts of levels and quality of education within the management committees (see Figure 6.3). Furthermore, education and management were discussed in the context of governance and independence of the cooperative movement.

Figure 6.3: Clusters of the relationships between themes, by word similarities

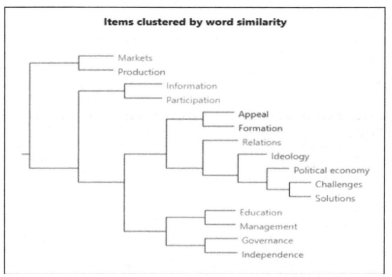

Source: Compiled by the author based on own survey data, 2018-19 (NVivo cluster analysis)

Each time an interviewee mentioned a problem, they also highlighted a solution to overcome that problem. These two, challenges and solutions were close to the political economy issues and the quality of ideology within the cooperative movement. I discussed the formation and cooperative appealing/attracting factor for people to join cooperatives issues. These results prove the relevancy, strength, and adequacy of the themes that I chose for this analysis. Markets and production and, information and participation, formed an isolated cluster.

Zimbabwe farmers' narratives: A genuine cry for independence

Following the methodology in Japan's qualitative data (see page 154), I identified a total of 15 sub-themes which I then reorganised into eight main themes from the interview and FGDs transcripts. These became the codes in which I organised text into as described in Chapter One, (page 18). These sections were as follows: i) Cooperative appeal, ii) Challenges, iii) Governance, iv) Ideology, v) Independence, vi) Management, vii) Political economy, and viii) Solutions.

Ideological backbone
Several scholars on cooperatives have given a great deal of attention to how cooperatives are managed, from the national perspective and individual cooperative level (Akwabi-Ameyaw, 1997; Birchall, 1997; M. Cook et al., 2009; Iliopoulos, 2017; Mariani, 1906). Moreover, in doing so, many have concluded, with less regard to the various other external factors, that it is in the management area that most cooperatives fail (Ortmann & King, 2007). While such lines of argument are plausible, other factors on the level and quality of management of cooperatives are equally essential. In this subsection, I examined whether the stakeholders of the cooperative valued the role that ideology plays in the movement. In addition to challenges in agricultural production (poor farming practices, imperfect markets, and climate effects), three other issues emanated from these interviews. The most significant problem has been farmers' attitudes towards the cooperative; then the level of homogeneity within the rural areas, and finally ideological contradictions within the government itself.

The level of cooperative education is not high enough for people to clearly understand how they fundamentally work. People need more information because as they get more information and more educated about how specific programs work, the more they will be

211

able to act as expected. Many of the farmers claim to know what the cooperative is, but only a few understand the fundamentals and principles of cooperatives. The co-operator we interviewed argued that:

[…] in the pre-independence movement, farmers had to understand cooperativism first. They were taught more on theory first, and then the practical aspects of the movement. This is what is lacking in all the cooperatives that came after independence and the ones that are operating now. However, again, joining a cooperative should be voluntary, even if the person joining it wants to enjoy a particular benefit, he or she should enter voluntarily. (CP Mathonsi Interview, 09/04/2018)

Many issues contribute to or affect the farmer's attitude and perspectives about the cooperative. The narrative of cooperative activities being a 'public good' is sometimes pushed by the politicians to link most rural government programs to the ruling party. Hence party allegiances become an issue in the distribution of public goods. In the absence of good education & information, members are highly susceptible to political manipulation. This challenge coupled with a disregard for cooperative principles affected the 1980s collectives' programs. Farmer attitudes based on misinformation can be argued to be the reason why most of the cooperatives formed by the government and those formed by NGOs failed. Once they feel that they do not have to put in any effort and that everything would be handed to them, then that cooperative would not last. Cooperatives demand sacrifice, dedication, and commitment from its members. Especially in the establishment stages. The CACU chairperson added:

I can tell you right now that most farmers who are in the cooperative think that cooperatives are government programmes that are meant to provide free inputs and free loans to the cooperative members. Farmer attitudes influence the success or failure of the movement, which is also seen through the harmful effects of the donor syndrome. Many donors offered many free goods and services through the cooperative. Many programs crumbled as soon as the donor pulled out, meaning that farmers attitudes are affecting the sustainability of the cooperatives. (CACU chairperson interview, 16/02/18)

Several respondents reported that the members of the cooperatives, especially in the resettled cooperatives, had too diverse farming objectives and goals (low levels of homogeneity of farmers). From these interviews, I learnt that a more homogenous group of people make the best sustainable cooperatives. The more problems or issues in common, the better the chances of the cooperative being productive and standing the test of time. Thus, without a central ideology that guides the overall objectives, the cooperative is useless.

I think the cooperative movement needs to have almost the same type of people faced with the same type of problems for it to be united and for it to work. The focus should neither be on whether one is a landholder nor on whether people who hold land are farmers. In this sense, there are many differences between people. So, each farmer's goals and objectives are too dissimilar to the rest. Smaller groups or cooperatives are a solution, like-minded people, together at first, and then eventually they become more prominent. F Mugaba interview, 08/02/2019)

One of the respondents highlighted that since ideology was unclear, the cooperative members did not know why they were in the cooperative.

To be honest, our cooperative is not doing anything right now (not productive). It was formed in order to access the tractors. We do not have any ideology, and we do not have one crop on which we focus. People are not working together to achieve aggregate goals. It is just a bunch of farmers who had their names written down in order to access the tractors from Brazil. Ideally, we are supposed to produce as individuals and sell as a group to cut costs, but this is not happening. People only collaborate and share implements. (F Manyika interview, 05/02/2019)

Another respondent added that before they took in more people into their cooperative, it was successfully run and had a clear ideology, goals, and objectives that the few members understood and adhered to:

> Before this (the tractor) cooperative came into existence we had another very successful one, it was called Back to Eden and had only eight members. It specialised in fruits and vegetables and had an ideology that encouraged vegetarian diets, hence the name Back to Eden. The merging of old & new committees was not smooth, and some of the incoming members did not trust the existing structures and wanted a complete overhaul of the committees. This caused many problems. Simply put, there were many differences in terms of ideology, trust and focus of the cooperative. (Mugaba interview, 08/02/2019)

Appeal of cooperative

Farmers, scholars, and policymakers are interested in cooperatives for several different reasons. However, a number of the people interviewed in the study highlighted that registration of cooperatives was a formalisation of everyday life because it is human to cooperate. For example, as one of the assistant registrars from the ministry reported:

> [...] cooperatives were actually started by the small-holder farmers so that they could acquire inputs in bulk, reduce transaction cost, and sometimes import inputs [...] lots of people are naturally into some collective action. There is a demand for cooperatives as can be seen through many people trying to register their groups. Although the number of unions is not increasing, the number of registered cooperatives is increasing, as seen through daily registration of cooperatives. (AR Mada interview, 08/03/18)

The biggest reason motivation for collective action in the rural areas, therefore, was to try and reduce the adverse effects of market imperfections such as high transaction cost and access to inputs and output markets. The message that the assistant registrar was trying to relay was more explicit in the words of the former cooperative registrar:

> Cooperatives, as we know them on paper, are just formalised reality. People cooperate in everyday activities, young man, you do know that even starting from Adam and Eve in the garden of Eden, they were cooperating. They probably should be called the pioneers of cooperation on this planet. The whole essence and meaning of life usually manifest itself as cooperation, tolerance and working together between human beings. (ER Mavhinyingwa interview, 17/03/2018)

Although the government has significantly reduced its financial support for the cooperative, it was still using the cooperative/group approach in almost all its rural programs, e.g., extension services. In addition to maintenance of standards, the cooperatives should improve farmers control of their produce and local economy by coming together and forming a massive pool of resources that is capable of lending money to the other various sectors. The same rationale was echoed by one of the members of the Tagarika irrigation cooperative:

> The cooperative was formed because of problems in accessing water around the farm. Since we had a dam nearby but no means of pumping the water to the fields, we had to come together to be able to buy the costly equipment. Irrigation equipment was too expensive to buy as individuals. Each member paid a joining fee of $100, and then we pay $10 every month afterwards. The benefits from the cooperative justified these amounts that we had to pay. (Tagarika FGDs, 31/01/18)

Farming on its own cannot sustain the lives of the farmers, but with cooperatives, the goal is to own the product, process it and add value to it. Farmers or people can influence government policy formulation processes when they have sufficient collective action:

> The farmers have no aggregate voice through which they can air their views. The rural areas are highly socially unorganised. These facts are also valid for peri-urban areas. There is so much potential for

215

cooperatives and so much scope to justify them. As we speak, I am involved in setting up a cooperative from below. The people in Beitbridge (where Mopani worms are grown) have for long been marginalised in terms of benefiting from the Mopani worm value chain. Formation of the cooperative will enable them to put a plug on income lost through surplus transfer as well as work together to establish sustainable ways of harvesting the worms and eventually developing a mechanism to increase their production. (Dr Matondi interview, 20/04/18)

There is a need to improve the level of beneficiation derived by the farmers from a bundle of available resources as in the Mopani worms' example. This process will be made easy by the establishment of an institution between the government and the companies, and it should overwhelmingly represent the wishes of the farmers. This institution should have experienced and qualified personnel in order to make sure that whatever program that was brought to the farmers has their concerns prioritised. In the next section, I discuss management issues from the national and the individual cooperative perspectives.

Management of cooperatives in Goromonzi district

From the government's point of view, there exist impediments to the smooth management of cooperatives. A higher number of unregistered cooperatives is one of the biggest challenges in the movement (depending on the economic sector). The agricultural sector has the highest proportion of unregistered cooperatives than in other sectors hence, making it harder for the state to administer the cooperative law in logistical terms. Part of the reason why the situation is so can be blamed on the government itself because registration of cooperatives can only be done in the capital city, Harare. The ministry knows it needs to increase support in the form of more personnel on the ground to interact with the farmers and train them. At the moment, a handful of staff in the department of cooperatives was too busy with clerical work and hardly had time to go into the field. Management improvements are necessary in terms of the cooperative register. I visited the Ministry several times to view

216

the Cooperative register (as is allowed by the CSA), but I could not get hold of this document.

According to assistant registrar Mada, differences in land sizes, tenure arrangements and products types complicate the smooth formation of cooperatives. This means that within the same sector of agriculture, management approaches for equipment, irrigation, crop, and livestock cooperatives differ significantly. While this speaks volumes about the need for the sector-specific legislature, it also highlights the fact that cooperatives should sometimes consider being product-specific, at least at the beginning. Lastly, the ministry officials also highlighted the negative effect of changing parent ministries every time the cabinet is reshuffled. Huge chunks of information, together with the concept of continuity, is lost with every change. The 'hand-over-take-over' processes are never smooth. They argued that it was one of the reasons why the cooperative register was not available.

Additionally, the unconducive macro-economic environment also affected the cooperative operation. Most cooperatives lost out during the dollarization of the economy in 2009. A number of them had huge savings, e.g., services cooperatives who had not invested in physical assets. The government sometimes import agricultural products at a cost higher than the cost of supporting the local farmers at the beginning of the season. Thus, some of the problems come from a lack of planning and prioritising.

As far as management issues at the cooperative level were concerned, lack of trust was at the top of the list. In an environment where corruption has been institutionalised in the government, private and civil society, cooperative members have grown suspicious of the top management. One of the leaders of Tagarika irrigation, which received irrigation and tractors on loan from the government lamented:

> Other challenges, especially within the cooperative movement and
> its management is that there are low levels of trust. However, it is not
> the fault of the cooperative leaders; for example, some farmers do not
> trust us because we could not tell them exactly how much we owe the
> government. We did not tell them because we also do not know. We

are still unclear about how the program actually works four years down the line. (Tagarika FGDs, 31/01/19)

In the FGD of cooperative chairpersons, solutions were solicited concerning the most significant management problem of unpaid or un-serviced loans. It was highlighted that the total CACU debt had risen to US$1.2 million and that if cooperatives failed to repay (through failure to collect the debts from their respective members), the property could be attached by the banks.

It is, therefore, the responsibility of the management to make sure that all farmers pay back their loans. The chairpersons here present (in the FGD) must come up with a plan to recover the loans. In a way, what I mean is that if the cooperative fails, it can be blamed entirely on the management, which consisted of the directors and the chairpersons here. We want any suggestion even if you think that it is silly or unhelpful. (Chairperson's FGDs, 19/02/18)

The FGDs opened a can of worms because many management malpractices were discovered. Some chairpersons habitually delegated their secretaries to write and reconcile their plan book as well as invoice books. Typically, the secretary of the cooperatives writes their separate reports, and the manager does the same. At the end of a given period, although written separately, the two reports should reconcile. This improves transparency and accountability in the cooperatives. However, if one person was doing them, then it defeated the whole purpose of accountability. This issue was noted by CACU officers who collect the books in order to check if they reflect the same transactions, and in more times than not, they were written in the same handwriting.

Since there is so much missing information in terms of who owes how much, the age of debt and the likes, the first thing to do after the meeting is for each chairperson and director to reconcile their books. Thus, membership books, membership fees, and debts had to be reconciled. Then the chairmen can go around collecting debts. (Chairperson's FGD, 19/02/18).

From the Xanadu A irrigation cooperative farmers interviews, I discovered that payment of cooperative subscription was a huge problem. Cooperative leaders were struggling to find ways to improve the payment of these fees, especially after the cooperation had become bigger.

> Our cooperative [formerly] had eight members that paid $10 per month as a subscription, which translated to $960 per annum. If we could pay this amount into the cooperative, it means we are productive. […] After forming a new cooperative, the new people do not pay subscription fees, and old ones stopped too. It is difficult to go after all the members. At the moment, farmers only pay for the services they hire from the cooperative, i.e., $100 per hectare for tillage, 60% of this money goes to repay our loan, and 40% goes into servicing of the tractors. (Interview with AR Mugaba, 08/02/19)

This is a severe problem because the people were 'forced' to form the tractor cooperative; thus, they do not see the need to pay subscription fees. The cooperative needs homogenous people, faced with similar problems, for it to be united and to work towards fulfilling its objectives. Thus, each farmer's goals and objectives are too dissimilar to each other. Smaller groups or cooperatives are a solution, like-minded people together at first and then eventually they become bigger. – (F Manyika interview, 05/02/19). In the same FGD, as a way to incentive management (since they are not paid), committee members suggested that members should be rewarded, by either giving them loans (if they applied) at little to no interest rates. Alternatively, they would automatically become beneficiaries of any cooperative programs. Thus, there should be provisions, either in the by-law or in this legislative tool, for incentives to the cooperative committee members. In the next section, we examine the role that the regulatory instrument plays in the development of the cooperative movement in Zimbabwe from the farmer's perspective.

Legislative instrument

The cooperative movement is governed by the 1990 act which was last updated/revised in 1996, and needless to say, many events have happened since then so that a revision was necessitated. Many cooperatives are not registered, and hence they cannot formalise their organisations/groups. There has been an increase in demand for cooperatives after the land reform (especially in the agricultural, housing and mining sector) which increased the pressure for the government to provide the cooperative services.

The farmers and the experts had divided opinion about the effect of FTLRP on the cooperative movement. Some of the respondents thought that it had increased the demand for cooperatives through an enlarged farmer base. Divergent views argued that the proliferation of the housing cooperatives (these types of cooperatives also increased after the reform) had almost destroyed the movement as a whole because they were 'fake cooperatives of convenience' (Interview with AR Nyamunda, 19/03/18). Interview participants also blamed inadequate legislative framework, which had many loopholes that had different consequences, and which corrupt individuals took advantage. For each economic sector, there exist several primary cooperatives, and all these are governed by one act: from mining, housing, tourism to agriculture. In this regard, they argued for separate cooperative laws for each sector.

There are three core structures of the cooperative, i.e., centralised, federated, and mixed. Zimbabwean movement seems to be federated. According to the act, the movement follows government administrative structures, i.e., the four-tier structure corresponds to the wards, 52 districts and ten provinces that Zimbabwe has. However, that is not the actual situation obtaining on the ground. The fact is there could be more than one secondary cooperative (focusing on the same type of produce) operating in the same district which is in clear violation of the act itself (Mavhinyingwa interview, 17/03/18). Similarly, there are different types of apex/union organisations in the same provinces focusing on the same cooperative business. There are also emerging complexities, especially in the newly resettled areas where farmers who were allocated land in the same 'former white-owned farm' are required to join the cooperative in that area. This emerging situation is not provisioned in the 1990s act. The CACU chairperson emphasised that:

The current act is also another source of headache. Zimbabwe has one umbrella Cooperative Societies Act of 1996. What is needed is a sector-specific act that is tailor-made for each economic sector, as in the case of the Japanese cooperative policy framework where you are doing your studies. There is no written criterion on who sits on the committee and how meetings are supposed to be carried out. The act, in its broadness, cannot address some of the sector-specific challenges. (Interview with CACU chairperson interview, 16/02/18)

An assistant registrar from the ministry echoed the same words:

[...] this necessitates a separate act for each cooperative sector since the issues that affect one sector cannot be applied to the other. In particular, I believe that the SACCOs cannot be administered effectively by the current statutory instrument. At least for now, the government should prioritise the production of a separate SACCOs law since this is a sensitive sector. (AR Nyamunda interview, 19/03/18)

The government is trying to follow the Kenyan model of establishing a robust SACCOs sector that can then power the cooperative movement. There is a draft law for the SACCOs under review, and a separate organisational body for SACCOs outside of the Cooperatives Federation body is also on the cards. This represents one step in the correct direction for the movement, but this should be the set-up for all other economic sector cooperatives.

Many of the expert interviewees argued that the current cooperative act was also detrimental because the government has abandoned the cooperative movement, specifically in terms of financial support. However, the act still made sure that the government maintained a firm hand on the movement, *albeit* using old and outdated parameters. For example, several advances had been made in understanding how rural areas worked, including the importance of social relations such as kinship in the movement (Chiweshe, 2011; Mafeje, 2003; Murisa, 2009). Yet these issues were not reflected in the approaches that the government used in administering the act. Additionally, the act still emphasised the superiority of choosing leader randomly with little regard to the members' skills and level of education. The CACU chairperson argued that:

> [...] the federations (even unions and Apex organisations) have poor leadership selection systems because leaders are chosen based on democracy with little regard to the qualification of the candidate. Even the act is silent on this issue. These then culminate in many management issues that undermine the movement [...] Again, to illustrate the fact that cooperatives are not a government priority, the constitution of Zimbabwe does not mention anything about cooperatives. Thus, only the act guarantees their existence. It should be enshrined in the constitution as in the case of Italy and other nations that did so. (CACU chairperson interview, 16/02/18)

Independence from the government

Several concerns were raised in terms of the amount of government influence in the movement. While a number of them were related to the legislative tool (as discussed above from page 219), other concerns took the types and manner of cooperative financing as well as the imposition of leaders that are sympathetic to the ruling party (as observed by Wedig & Weigratz, 2018 in the Ugandan cooperative movement). The sources of funding for social movements across the world have always had defining consequences on the eventual ideology and general trajectory that a cooperative took (Dogarawa, 2010). The basic principle is for every cooperative to be self-sustaining; this way, it can choose its objective and make decisions that genuinely benefit the members. The Zimbabwe government funds the cooperatives, and the Ministry has the power to appoint all the committee members of the national cooperative central fund. This is a bi-pronged control system (Interview with AR Mada). Additionally, NGOs have played a significant role as well in terms of financing and creation of cooperatives. Therefore, cooperatives need to eventually move towards self-financing as exemplified by the Japanese cooperative movement. An assistant registrar added that:

Most of the cooperatives on the ground are too political. Either they were formed by the state (top-down approach) or the management committee deliberately engaged themselves with politics in the hope of getting favours from the political elites. For most, it is about forming rallying points and nothing about uplifting the standard of living of the farmers. The solution to this is to encourage grassroots cooperatives to be formed. (AR Nyamunda interview, 19/03/18)

The co-operator weighed in on this subject by adding that:

I firmly believe that the government needs to stop giving farmers free inputs through subsidies because farmers always develop dependency syndrome. If farmers form cooperatives, then the government can lend money to these people through the cooperative. (CP Mathonsi interview, 09/04/18)

Another contribution came from the scholar:

One of the weaknesses is that the movement and its members heavily depend on the government for financial, technical, legal support. This is OK [in the initial stages], but the government needs to wean-off the farmers gradually. (Dr Matondi interview, 20/04/18)

On the contrary, the reduction of persons in the ministry who are in charge of cooperatives can be a positive development for the cooperative, only under the condition that the Zimbabwe National Cooperatives Federation (ZNCF) is properly constituted, functional and independent from the state. In such a case, the ZNCF would quickly take up functions such as auditing and training for their member cooperatives. As it stands, the federation only has two members from CACU, and the rest are from the housing cooperatives. That means from a total of eight economic sectors in Zimbabwe, only two sectors are represented in the federation, and thus, it is improperly constituted/unrepresentative (ER Mavhinyingwa interview, 17/03/18). Properly constituting the federation will ensure that the federation takes over from the state; training, educating, providing recommendations for the by-laws

223

while state functions would be limited to regulation and registration of cooperatives. According to the former registrar, the cooperatives should not get any special protection from the government through free inputs, but it should be supported so that it can be competitive on open markets. The Tagarika FGD agreed that the role of the government is essential and cannot be done away with:

> In this respect, the state should be involved more in the cooperative because it is the interest of the government for the cooperative to prosper. However, they need to keep giving advice and not to command us. Most importantly, it should listen to the people on the ground who experience everyday farming challenges. (Tagarika FGDs, 31/01/19)

One of the leaders from the Tagarika irrigation cooperative spoke about the nature of the tractor cooperative which they had entered. He summarised the problem of the attitudes among farmers which we discussed earlier (page 216). Cooperative ideology and sound leadership should avoid such kind of programs in which free things are given to the cooperative. Although the nature of the Brazil tractor cooperatives remains open to research, cooperatives should be careful in terms of the programs that they sign up. The state should also protect the farmers by perusing, at the national level, such programs as the tractor cooperative scheme and some of the Sino-Africa deals. Some deals at the national level might lead to possible appropriating portions of their land or resources as feared China-Africa researchers. In the case of cooperatives, by receiving free inputs and equipment, they become more susceptible to political manipulation. In their words, they said:

> Of course, we will not be able to pay even a quarter of the cost [of the tractors], I did the maths, what we pay back as debt servicing is nothing compared to the cost of the cooperative tractors and other equipment. So, in a way, these were free. Moreover, the reason why we have to pay is so that they [the government] monitor tractor servicing. It is a way to ensure sustainable use of tractors. This move is good and is different from Gono's [Former Governor of the RBZ]

224

mechanisation program in which we knew that the inputs were completely free. (Tagarika FGDs, 31/01/19)

Other independence issues emanated from the power that the cooperative registrar held within the movement. They are responsible for registering, training and arbitration of disputes within the cooperatives. During the formation, they provide the cooperative with a template of the by-laws, which the cooperative members adopt as it is. The former registrar echoed these words:

Cooperatives are non-autonomous. For example, the government is supposed to help them in drafting their cooperative by-laws; however, the prospective cooperatives adopt the template by-laws that are recommended by the ministry almost as they are. The worst part is that, after adopting these recommendations, the majority of the members never try to understand them. In the same sense, the government does not carry out enough training workshops at the member level to educate the people on what the by-laws mean. (ER Mavhinyingwa interview, 17/03/18)

The political economy of cooperative activities in Goromonzi

It is essential, as I describe the contemporary Zimbabwe agrarian situation, to explain the political economy of cooperative development at local level. This will help delineate the role of the government in cooperative development. Ideally, the objective of the government should align with that of cooperatives (which is rural development and uplifting of standards of living). However, government leaders are politicians, and politicians make political decisions that sometimes diverge from economic, cooperative and development objectives.

In the context of looking at the broader issue of failed strategies of economic development throughout Africa, many analysts now argue that the central reflective feature to consider is the political economy that surrounds development policies formulated and implemented since post-colonial days. (Akwabi-Ameyaw, 1997, p. 440)

A cooperative movement funded by the government is susceptible to political manipulation, whether there is a strong ideology or an ideal legislative instrument, as discussed on page 211. I expand this discussion in this sub-section and seek to decipher the political economy of the movement. Again, as in the previous subsections, I analysed these issues from the government, the opinion leaders, and the cooperative members' perspectives.

The interviewees had a differing understanding of the trajectory of cooperative government policy over the past 40 years. While some understood and defended the government's role, others chastised it through some interesting *'conspiracy theories'*. For example, it is believed that not only did the government extend its arm into the movement through the statutory instrument, but it trained and planted *'their people'* into various critical leadership positions:

> There was too much interference from the state as it had too many politically affiliated people running the cooperatives, especially in the housing sector. The state imposes leaders within crucial decision-making position; for example, no one knows where the five leaders from the housing sector in the federation came from. This is detrimental to the movement. The agricultural sector seems to be abandoned at the moment with movement only visible in the housing. (CP Mathonsi interview, 09/04/18)

Mr CP Mathonsi went on to talk about the role that former Zimbabwe Independent newspaper columnist and economist Eric Bloch (1939 - 2014) with the help of John Robertson (economist) had on the Zimbabwe dollar ever since independence. He was emotional as he spoke about this issue. He argued that the economists advised the Reserve Bank of Zimbabwe to devalue its currency arguing that it was always overvalued. Devaluation always happened towards the end of the farming seasons, i.e., the start of the marketing season for farmers. The effect of devaluing the dollar undervalued agricultural produce since it would be sold at relatively lower prices. Produce buyers (who were international middlemen with links to the European markets) made a killing from the transactions. In this respect, the government either did not know or

226

were also interested in underpaying the farmers for their produce. He firmly believed that if the cooperative movement were strong enough, it would have been able to pick this and put a stop to it or benefit from it. The same line of thinking was held by the scholar:

[…] but then again, weaning the farmers so that they can stand on their own is not always a government priority. For example, tobacco may be listed as one of the primary foreign currency earners after mining. This means the state is also interested in getting their hands on the part of the US$600-900 million in foreign currency that comes from tobacco exports. In such a case, the state will not act in any way that would compromise the buyers of the tobacco. A lot of the times, the state always protects the wishes of the foreign companies ahead of the peasantry. (Dr Matondi interview, 20/04/18)

Other decisions seem to defy economic or development logic. I did not get satisfactory answers from the ministry officials in terms of the motivation for keeping the old CSA despite two better revisions proposals in 2005 and 2017. I was inclined to postulate that these decisions were political calculations that placed specific individuals at a better political vantage point than the other. The former registrar said that he is still trying to figure out the reason why he was 'forced to retire early' from the Minister of Youth Development and Employment Creation (MYDEC) cooperative department.

I firmly believed that the momentum that my team and I had about cooperatives in 2000 was high enough to produce positive development outcomes if it had been allowed to continue. I, therefore, blame the former Minister of Youth Development, Gender and Employment, Border Gezi for killing the momentum when he forced all the top management in the cooperative department to retire. However, together with my forced retirement (at the age 57) went with the prospects of the cooperative development policy which was supposed to inform a new CSA. The policy that came in 2005 had been put forward by Moven Mahachi before his death, and they had only revised it. Ambrose Mutinhiri later published this proposal for cooperative development policy in the Ministry of youth development, but it was never implemented. (ER Mavhinyingwa, 17/03/18)

As noted earlier (page 230), there is also another proposed cooperative development policy of 2017 that was on the cards. However, just like many things in Zimbabwe, the November 2017 events, and the July 2018 elections affected the momentum of this cooperative policy reform. Parent ministries changed, the SMECD was disbanded, and the cooperative department was placed under Women Affairs ministry. The Cooperative Societies Act of 1996 needs to be amended to bring it in line with contemporary challenges brought about by the land reform, the changing political landscape as well as the 'Zimbabwe is open for business' stance taken by the government.

In addition to the contradictions that existed during the time cooperatives and small-to-medium enterprises were run in the same ministry, there seemed to be some confusion in the ministry as to who was the exact cooperative registrar at the time the interviews were carried out. The most probable explanation is extreme power struggles within the ministry or the department. In most cases, people fight over positions because they want to be in charge of programs, thus increase the chances of abusing privileges and personal gain. These add complexities to the political economy of cooperatives from the government level.

While the narrative in this book so far has, to some extent, painted cooperatives leaders and their members as mere victims of government politics and manipulation, it is essential to note that they have agency too. Cooperators get free inputs from the government or the NGOs, and they know that these free inputs will always come. Thus, if they realize an opportunity of getting free inputs or equipment, and that all they need to do is to form a pseudo cooperative to extract benefits, then they will do it. Although this phenomenon is also present in the agricultural sector, it reached unhealthy levels in the housing cooperative sector when urban land grabbers took peri-urban land and created cooperatives in order to register and develop the occupied land. An assistant registrar had an opinion on the issue:

Many cooperatives, especially the housing cooperatives, were formed after grabbing a resource. Thus, they were motivated by the need to legitimise their resource grabbing. This violates some cooperative principles, mostly the need for voluntary membership. The cooperative movement is being misused and abused. The members are not genuine, but they are opportunist, and they want to exploit and accumulate resources for themselves. The sooner we acknowledge this fact, the sooner we reconfigure the direction of the cooperative movement. (Interview with AR Nyamunda, 19/03/18)

The management committees deliberately engaged themselves with politics in the hope of getting favours from the political elites. However, this led to corruption which is perhaps the biggest impediments to the smooth collection of cooperative debt. Some cooperative chairpersons were using politics and political muscles to take fertilisers from cooperative programs. In most cases, these chairmen re-sold the inputs, were not productive and never paid back. Also, they were widespread favouritism, in which chairpersons allocated fertilisers to their relatives that were not even cooperative members. CACU itself recorded these gross violations, and the members used political connections to evade CACU and the arm of the law. Some extreme cases were reported in which fake farmer names were used in getting fertilisers from the CACU. These politically connected people would then intimidate and threaten the agricultural extension officers and the other cooperative members with untold harm if they dared to pursue the issues further (Chairperson's FGDs, 19/02/18).

However, about three-quarters of the meeting (this was in 2018) agreed that these individuals do not have this power anymore ever since the new dispensation came into power in November 2017. The government of Emmerson Mnangagwa had managed to remove some of the power that these chairmen abused because they were aligned to the late former President RG Mugabe. The local authority (elected councillors) also had an important role to play in dispute resolution in the countryside. Additionally, engagement with traditional authority (chiefs, headmen and village) was another way of drumming support for the chairmen to do their jobs without fear.

However, precaution should be exercised since the administrative authority is riddled with politics of its own (Chairperson's FGDs, 19/02/18). A member of the Xanadu cooperative spoke about the politics and cooperatives:

> Although it is tough to do, cooperatives need to be apolitical since most of the politicians are corrupt. Thus, if they enter into politics, then they will become corrupt as well. Also, once people start doing politics, it means people from party B cannot join the cooperative if the leader is supporting party A. (F Manyika interview, 05/02/19)

However, three years into his term, new widespread mass media reports claim that R.G Mugabe affiliated corrupt cooperators have been replaced by those affiliated E.D Mnangagwa. The scandals in the Command Agriculture scheme seem to confirm the continuation of the Mugabe status quo.

Looking the other way: Profiling the attempts to revise the cooperative policy.

To understand the role of the state in Zimbabwe's cooperatives, I interrogated; *1)* the Cooperative Societies Act of 1996 together with the other supporting legislative instruments; *2)* the evidence from various in-depth interviews carried out with key informants, ministry officials, experts and cooperators (cooperative practitioners) and *3)* the farmer attitudes about the nature and extent of state hegemony in the cooperative movement.

The Cooperatives Societies Act of 1996 (Chapter 24:05) (CSA henceforth) was put in place to provide for the formation, registration, management, functioning and dissolution of cooperative societies across the eight different sectors of the economy. Here lies the first legislative problem. Most countries have sector-specific societies acts, while others have two or three laws within each sector. The Zimbabwe movement has only one for the different types of cooperatives in the economy. Although some scholars argue that one blanket law is more comfortable to administer, the problems of fake housing cooperatives in the Zimbabwe movement poke holes into that argument (Mavhinyingwa interview, 17/03/2018). My analysis of the CSA picked some critical provisions from the Cooperative Societies Act of 1996 that require immediate amendment. These have conceptual, ideological, and practical implications for the development of the cooperative movement.

Throughout the discussion so far, I have consistently argued that the government has too much power within the cooperative movement. This is not my opinion only but by other scholars (Makochekamwa, 2015; Misininga, 1988), co-operators (as shall be discussed in the next sub-sections) and even the government of Zimbabwe itself (MYDEC, 2005). I come to this conclusion based on the proposed revision to the 1990s Cooperative Society Act which was done by the ministry itself. The first proposal titled "The revised Government policy on Cooperative development" was in 2005 under the guidance of the then Minister of Youth Development and Employment Creation (MYDEC), Cde. Ambrose Mutinhiri. In general, the propositions in this revision were too radical from a government point of view but would have helped the movement. Since most of these propositions were moving ultimate power to the cooperative movement and away from the state itself, it is not surprising that the proposal never saw the light of day.

The document noted that by 1987, there were 1,800 cooperatives which rose to 3,575 (six Apex and one federation) by 2003. An estimate made in 2003 put the total membership approximately around 200,000 nationwide. The number of members keeps growing, given that by 2017, the movement had 465,000 members. The policy document highlighted how the Savings and Credit Cooperative (SACCO) and the housing cooperatives spearheaded the current cooperative movement. The SACCOs had managed to mobilise savings of approximately US$675 million from a share capital of US$40million and a total loan portfolio of US$500 million, that was in circulation within the movement (as at 2003).

Interestingly, the cooperative policy identified and acknowledged that the government had changed its political ideology from socialist to market-oriented, considering the fall of socialism on the international platform. With the change in philosophy came a considerable loss of qualified and competent staff in the movement, which had catastrophically reduced the quality of support from the government side. This admission underscores the fact that the personnel within the government is not skilled and lacked the experience necessary to transform the movement.

Furthermore, the report also confesses that most cooperatives formed after independence were mainly state and donor cooperatives. The members never identified with the movement either through an investment of equity, commitment, or dedication. So, as highlighted in Wedig & Weigratz (2018), the movement was seen as a state or donor movement in which the farmers only had to register to benefit from the initial capital donations. This fact profoundly affected the sustainability of the cooperatives.

The revised policy document has a lot of positive proposals which if followed, could revolutionise the movement. The ministry acknowledged its excessive power through the registrar which enabled 'the government to exercise too much control in the administration of the cooperative, thus inhibiting their growth as independent business organisations' (MYDEC, 2005, p. 2). The regulatory environment since independence (and even now) is highly restrictive and prescriptive and hence cannot oversee to genuine cooperative development.

Attempts to change the direction: New aims and objectives.

The suggested objectives and aims would have genuinely changed the direction of the cooperative. I argue that this instrument should have been implemented. It sought to move away from the non-viable and highly unsustainable cooperatives that depend on donor and state funding; by fostering the grassroots capacity building of the members so they could empower, govern and manage themselves. The specific objectives are as follows:

- To create a conducive environment for the growth of the cooperative movement in Zimbabwe
- To address the new challenges faced by the cooperative movement since the pronunciation of the 1984 cooperative policy. These challenges include poverty eradication and employment creation, globalisation, changes in the economic orientation,

development of the SMEs, the agrarian reform, and the indigenisation of the economy.

• To realign the government policy on the cooperative development with the ILO recommendation 193[18] on the promotion of cooperatives adopted in June 2002 (sic) (MYDEC, 2005, p. 3).

In achieving these seemingly progressive objectives, I argue that in addition to the problems brought about by neoliberalism in the countryside, the other closely related big challenge to cooperative development is the building of the small-to-medium enterprises (SMEs). This situation led to policy and ideological contradictions because the government tried to simultaneously develop cooperatives and the SME by administering them through one ministry (2013-2018). SMEs are small private-owned companies with a minimal number of workers, capitalisation levels and turnover ratios (Mavhinyingwa interview, 17/03/2018). Thus, in the sense of the rural areas, individual farmers can be described as having small plots, hiring very little labour and equally smaller turnover ratios. In this sense, they become agricultural SMEs. Hence having a ministry that is responsible for promoting SMEs at the same time as promoting cooperatives in the context of the rural areas was self-defeating because the aim is to get these farmers into cooperatives and not deal with them as separate SME entities.

Although the policy tried to link the two, it described the SME as a cooperative type of business which is mostly untrue. SMEs are a direct competition to the cooperative movement. To a greater extent, the government's stance on SMEs is to actively develop an indigenous local black capitalist class, which is not what cooperatives do. Although cooperatives can make a profit, their primary objective is to improve the lives of the members. The government should provide the conducive environment through legislative support as well as periodic training for the farmers to stay abreast of the latest developments in business management skills. In this proposed policy,

---

[18] "The ILO recommendation 193 is the only international policy framework for co-operative development that has the added value of being adopted by governments, employers organisations and trade unions, and supported by relevant civil society organisations" (Smith, 2004, p. 8)

the government wanted to wean off the cooperatives and allow for a new wave of cooperatives to take precedence.

Proposed source of funding and the roles of the private sector.

Access to funding is one of the biggest hurdles to cooperative growth. The government proposed to encourage members to be more proactive and to fund their economic ventures through joining fees and capital contributions. Such possibilities are, however, slim since the peasantry has so many poor farmers who cannot afford to raise enough money for joining fees. Nonetheless, Chayanov (1991) highlighted that the peasantry is not homogenous, and some farmers can raise money for joining fees, exhibit self-reliance through share-capital contributions, and eventually generate reserves still government input is required.

Since cooperatives are encouraged to become more autonomous, revisions in the Act proposed for cooperatives to establish self-help financing mechanisms. One of the channels that they can take is borrowing from banks or government institutions. However, clear good debt repayment history must support borrowing. As a response, the government then will encourage the formation of a unit within the cooperative movement that will be responsible for the human resource development and provision of consultation and advisory services as well as auditing and supervisory services to its members. If the government had adopted this (14 years ago), the cooperative movement in Zimbabwe would be at a better stage at the moment.

The 2005 revisions had also acknowledged the differences the CSA and the ICA principles and had proposed to update them. The policy also addressed the need for the movement to be in line with the rest of the world and be on the same level through the ILO recommendation 193. The delay and disregard of this incredible piece of legislation can be explained by the complexity of the political economy of cooperative development in Zimbabwe. The cooperative movement is considered a tool by some elites used for gaining support in the rural areas or accumulating through corruption in the urban housing cooperative. This instrument was also slightly revised in 2017 but has also been ignored for a second time.

The relationship between government and cooperative enterprises

Cooperatives do not get any special treatment from the government anymore, especially in terms of finance. However, neither of the ministry officials nor the leaders of the cooperative movement were willing to admit this reality. Therefore, it may mean that most of the co-operators were not aware or had not realised the shift in government policy. The policy revision was supposed to clarify this by highlighting that the cooperatives need to quickly reorganise themselves, implement organisational reform programs that enable them to manage and control their affairs. It categorically states that the government would no longer have any obligation to protect or support cooperatives any more or less than the private business enterprise. The state would only train and build capacity with some minimum necessary regulations reformed slowly. This process is a huge game-changer that should have sent shockwaves throughout the movement. The rationale is to rightly reduce the footprint of the government in the cooperative movement.

From independence, the government has been acting as '[…] the lawmaker, the promoter, the supervisor, the service provider, the funder and the arbitrator of disputes […]" (MYDEC, 2005, p. 16). Moreover, this has undermined cooperative growth since the state has interfered too much in cooperatives affairs (what Scott – 1998 called a high-modernist approach). Therefore, within the confines of the proposed revision, the state wanted to limit its role in enforcing the act. Policy frameworks, especially concerning the control of food trade and other regulations, is an issue of sovereignty and national vision, hence state role should not be wholly eliminated. As we saw, in the case of Japan, the JA progressed through powerful lobbying activities to Liberal Democratic Party administration in the 1960s. And in the case of Southern African countries after independence, food producers were also protected by the state-controlled marketing boards. My view is that the time of protectionism may almost be over. Still, the key to achieving positive development outcomes is a gradual disengagement of the state while maintaining some key regulations. The abrupt halt in support as done during the ESAP may do more harm than the maintenance of the current status quo.

The development will also depend on the level of communication and information sharing between the government, the private sector, and the cooperative movement. For instance, the revision proposed the government to relinquish more power to the hands of the movement, such as auditing of primary cooperatives. Proper structures need to be put in place first to enable smooth design and implementation of comprehensive cooperative re-organisation programs which are in line with national socio-economic development policies. Without communication, assistance, funding and training from the government itself, this task possesses severe challenges for the cooperative movement.

The 2005 document also brought into perspective, the reality that increased numbers of cooperative transcend to increased demand for cooperative training and skills development. At the same time, cooperative funding for training, skills and education from the government has declined in the past three decades. Thus, the government is currently unable to channel resources to training and education. The policy further proposed that any training by the government will no longer be free of charge but will be charged based on cost recovery.

As a way of lessening the burden on cooperatives, the policy proposed that the government incentivise the cooperative through tax cuts, duty-free imports of equipment, tax holidays, non-taxable income as well as being given some tenders on some public works programme, especially in the rural areas. However, this is to the discretion of the government, and these incentives will be reviewed annually. My assessment of the proposed policy revision is that it was a step in the right direction. Policymakers need to view cooperative development as less of decentralisation of power or creation of more robust, educated peasantry who may rise if their needs are not met. Nevertheless, the state should view them as a group of people who can take themselves out of poverty through self-help. The next sub-section examines farmers perception about the government's role in the development of the cooperative movement.

Farmers attitudes towards state cooperation

Data is primarily based on field interviews and observations. I asked the cooperative members whether they had ever met any government officials responsible for cooperatives, and how many ministries they had met. Some farmers (38.5%) highlighted that they only encountered with the government or its apparatus during the formative stages of the cooperative. This is the time when the cooperative applies to the government to become a registered cooperative. Approximately 20% agreed that they had met government officials, not necessarily the cooperative registrar nor the deputy registrars, but members of the ministry. These officials had come to their areas with a training and education program (20.8%) or were disseminating information (20.3%). Other farmers pointed out that some officials came to their area to check their accounting books (16.7%) and help/monitor their financial wellbeing (15.1%).

Approximately 15.1% of the respondents had highlighted that they had encountered with ministry officials when they came for dispute resolution within the cooperative movement. The government is mandated to carry out these functions by the Cooperative Societies Act of 1996. Only 7.1% and 14.3% of the farmers in Xanadu A and Shungu Organic farmers had seen or encountered with government officials. The rest of the cooperatives had higher proportions of farmers that had seen government officials in their area, Chikwaka dairy (65.5%), Gosha eggs (47.8%), Gutu golden egg (58.3%). The cooperative that had witnessed the highest number of visits from government officials was Tagarika irrigation cooperative (76%) of the farmers. While other cooperative members had seen an average of two ministry officials, members of the cooperative in Tagarika had seen and known an average of seven government representatives who frequented their cooperative. After analysing the data, I revisited the cooperative to find out the reason for this exceptionally high level of contact with the state. I found out that the cooperative is a recipient of tractors from the government, and it is the nature of the tractor agreement that officials should come to the cooperative headquarters regularly.

These results reveal how the state is selective in contacting cooperatives. Its contact level depends on the size of the cooperative and the importance of the crop that a cooperative was producing. Relations also depended on whether there are any on-going joint projects between them and the cooperative. All this has implications on the nature and character of the cooperative that would be viable in Zimbabwe's agrarian structure. For example, based on this result, organic vegetable producing cooperatives would get less attention as compared to a tobacco-producing cooperative since tobacco is an 'essential crop' to the country (as a principal foreign currency earner). In the same instance, smaller cooperative struggling to invest in machinery would get less attention than a prominent cooperative that has been able to accumulate assets and infrastructure. In the political-economic analysis, the state is interested in what it can get from the cooperative movement before what it can give.

Although the state is not as present on the ground as it was in the 1980s (especially during the collectivisation period in the model B resettlement areas), the farmers still maintained that they always feel that the government is watching their activities and monitoring them. A third reported that they were undecided on whether the government had a heavy hand in the movement; however, an overwhelming 63% agreed that the state was omnipresent in the movement (*Table 6.2*). As expected, all the members from Tagarika cooperative (which experienced the most visits from the government) agreed to this assertion.

There were mixed feelings about whether the government should increase or cease its involvement in the cooperative movement among all the cooperatives. Some cooperatives such as Gutu eggs and Simba Ivhu cooperatives were of the notion that the government need not be involved with the cooperative movement than it already was. The rest of the cooperatives thought that there would be no harm if the government increased their role. Of note is Xanadu A and again, Tagarika cooperatives, which recorded the highest proportions of farmers (88.1% and 62% respectively) who agreed that the state should be involved more in the cooperative movement.

Table 6.2: Attitudes towards state hegemony in the cooperative movement

	Response	1	2	3	4	5	6	7	8	9	Total
a) Ministry has a heavy presence in the operations of cooperatives	Strongly agree	3.4	34.8	0	0	14.3	11.1	7.7	6	19	12
	Agree	48.3	17.4	0	42.9	14.3	11.1	23.1	92	61.9	51
	Undecided	41.4	47.8	100	57.1	71.4	77.8	69.2	0	9.5	33.3
	Disagree	0	0	0	0	0	0	0	2	7.1	2.1
	Strongly disagree	6.9	0	0	0	0	0	0	0	2.4	1.6
b) The ministry should be involved more in the Cooperative management	Strongly agree	0	0	0	0	0	0	0	14	85.7	22.4
	Agree	27.6	0	0	0	57.1	0	30.8	48	2.4	21.4
	Undecided	62.1	82.6	0	85.7	14.3	22.2	38.5	36	4.8	37
	Disagree	10.3	17.4	100	14.3	14.3	66.7	30.8	2	4.8	17.7
	Strongly disagree	0	0	0	0	14.3	11.1	0	0	2.4	1.6

1=Chikwaka, 2= Gosha, 3=Gutu, 4=Kumboedza, 5=Shungu, 6=Simba Ivhu, 7=Survival skills, 8=Tagarika, 9= Xanadu A

Source: Compiled by the author based on own survey data, 2018-19

Interviews highlighted that since these cooperatives were recently established (Tagarika in 2014 and Xanadu in 2015), the farmers felt that they needed more government's help (Xanadu & Tagarika FGDs, 2019). Irrigation equipment and tractors are expensive and given the few options for farmers to obtain credit; the farmers expected the state to take care of its farmers in their formative years. These findings imply that the cooperative model that I should come out with should factor in formative years of the cooperative and how 'teething problems' can be overcome. Decentralisation can be a solution to this dilemma as seen in countries such as Nigeria, Japan and USA. The presence of the national government in the local spheres is only as pronounced as the prefectural government wants it to be in these countries. Financial, legal and administration coordination remains with the central government of course. Still, day to day issues such as extension and agricultural training go to the local government who are well in touch with the local realities. This can be one way of reducing information asymmetries.

Contemporary issues in Zimbabwe agriculture

Throughout this chapter, I have highlighted several challenges that affect the farmers and the movement in passing. However, it is necessary to dedicate a sub-section to this discussion. I asked the farmers to rank the challenges they faced; 1) as farmers trying to access resources necessary for agricultural and social production, 2) as farmers trying to offload their produce into the commodity markets, and 3) as cooperative members involved in all matters that concern the movement.

Agricultural financing and other agricultural challenges
A higher proportion of the farmers ranked access to inputs as the most significant challenge. They believed that if they could access inputs, then all other problems could simultaneously and eventually be resolved. In theory, the cooperative is supposed to take care of this. It should be able to mobilise funding and identify cheaper options for the farmers to buy inputs and cost-effective transport to rural areas. However, Zimbabwean cooperatives were focused on selling output and hence had neglected their role as a farm supply and marketing organisation. The movement needs to reclaim its role as the supplier of inputs, especially in these times when the farmers have limited affordable options. Lack of working capital, droughts, and lack of access to credit were ranked second, third and fourth, respectively. This is problem affects not only farmers in the cooperative, but it is a problem that affects all farmers across the nation, and even affects other economic sectors (World Bank, 2019a). In order to achieve agricultural development, attention will need to go to the provision of credit and funding to agriculture.

Zimbabwe has periodic low rainfall (and high temperatures) patterns occurring in NRIV & NRV every five years and across the whole country every ten years (World Bank, 2019a, p. 16). The advent of climate change has thus taken its toll on Zimbabwe as well. Before the 2000s, the whole country used to experience severe droughts once every decade, but recently they have been more frequent and are pronounced by mid-season droughts (World Bank, 2019a, p. 2). More irrigation cooperatives such as Tagarika cooperative can be the

240

solution in this case. Over 75% of the farmers reported that they had lost their property due to theft and that it was indeed an issue of concern. In most cases, those who steal do so in order to feed their families because they do not have stable livelihoods. It is also an area that cooperatives can play an essential role since it diversifies livelihood options for the farmers.

High transport cost, animal disturbances, high worker wages, access to markets and labour shortages were all reported to be challenges by 40 to 50% of the interviewed households. The cooperative can resolve these problems by reasserting its input supply component of the 'supply and marketing cooperative'. Other problems such as ageing farmers, small land size and land tenure insecurity ranked lower.

In my study, a tiny proportion of the farmers were able to lease-in land from other farmers. Hence few people reported land shortages as a problem. The same can be said for land tenure insecurity (*Figure 6.4*), meaning farmers feel very secure about the ownership status that they hold under the current land administration system.

Figure 6.4: Challenges and constraints faced by farmers during agricultural production.

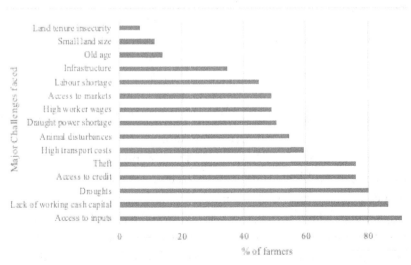

Source: Compiled by the author based on own survey data, 2018-19

In this respect, farmers need to pressure the government and the private sector some more, so that banks can start accepting the current tenure documents as collateral security to obtain loans. A lot of the pressure necessary to achieve this can be mobilised or gathered through the cooperative system. Thus, we continue to see the relevance of the cooperative system in Zimbabwe.

From my data, the middle-rich cluster had the highest proportion of farmers (90.3%) who experienced constraints during marketing time. There were no significant differences in terms of the proportion of the farmers who had issues during the marketing of their crops across the remaining peasant clusters. Overall, 47.4% of all the respondents had experienced challenges in marketing their crops and livestock between 2015-2018 seasons (*Figure 6.4*).

Further analysis unpacked the nature of these marketing issues since my data demonstrated that the cooperative has abandoned their supply mandate and was instead focusing more on commodity marketing. Hence, there are not supposed to be any challenges in marketing. *Figure 6.5(a)* illustrates that the bulk of the people who faced challenges in marketing their produce did so in the maize crop (67%), implying that the challenges were specific to a particular grown crop.

It also means that, given that none of the cooperatives focused on the production of maize *per se*, and thus the cooperative was not responsible for its distribution. In fact, maize formed part of the 'agricultural production outside cooperative structures'. The second type of crops that had challenges were all horticultural, but these only accounted for 23.1% of the farmers, while soybean (another non-cooperatively produced crop) had 14.3% of the producers complaining. *Figure 6.5*(b) shows the type of challenges that farmers experienced, especially for maize marketing.

Approximately 73.3% and 68.1% of the farmers that had experienced problems during the marketing of their crops reported reduced prices and lack of transport services as the top issues, respectively. The proportion of farmers that reported cost of handling the output as too high and hence, reduced their net income levels then followed. Other challenges such as unavailability of commodity buyers (markets), inadequate price information and theft were also reported, *alas* by smaller proportions of farmers. What can be gleaned from these results is the fact that the cooperative system could resolve most of the significant challenges faced in the selling of commodities. If cooperatives could be developed for maize production as well, then these issues would become a thing of the past.

Figure 6.5: Agricultural commodities and the marketing challenges faced (2015-2018)

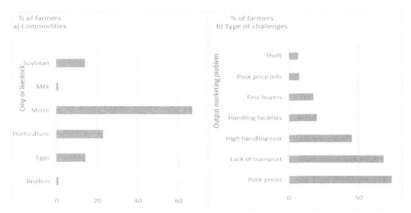

Source: Compiled by the author based on own survey data, 2018-19

Challenges faced in the movement.

The nature of challenges in the cooperative movement is diverse. A concise understanding of the problems in the movement is necessary for the construction of a cooperative model. To achieve this, I firstly present evidence based on an emerging problem in the Zimbabwe cooperative movement. In light of reduced government funding in the cooperative sector, a new wave of cooperatives is emerging in the newly resettled areas. However, the government remains a threat to the genuine development of these cooperatives.

The challenge emerges in three different but related ways. The first one is 'intentional' maintenance of an old Cooperative Societies Act (1996), which is out of touch with reality (they had chances to reform the act, but they have not done so up to today). Secondly, the government is 'unintentionally' destroying the cooperative movement by feeding them with free inputs and implements again, repeating the same mistake of the pre-2000 agricultural cooperatives. Thirdly, the government is still using the same hubristic approach to rural development with little intimate knowledge of the socio-economic issues underlining emerging cooperatives in rural areas.

To illustrate this, I used a case study of the two cooperatives in the resettled areas in Goromonzi. For *Textbox 6.1*, I asked the leaders of the cooperative to describe the development of their cooperative. From their narrative, it was clear how the presence of the government became an obstacle to cooperative development.

The problem is multifaceted. On the surface, it may seem like the problem of the government policy, but it also has a lot to do with the farmers' mentality or attitude towards government programs as alluded to earlier (page 209). Firstly, even though the intentions of the government were good, they did not sit down with the farmers or the already existing cooperative to try and see how the tractor program could be integrated. Secondly, on the side of the farmers, before the introduction of the program, people knew that they had to work hard for the cooperatives to work and that the money that they had invested in the form of subscriptions had to bear dividends (*Textbox 6.1*). However, as soon as the government brought in a program, even some of the eight members who used to keep up to date subscription accounts failed to do so. The mentality that this was a free program (in which they did not have to pay anything) affected their attitudes.

Additionally, the political leaders who are sometimes opinion leaders in the rural area advertise these programs as free government programs to benefit the political followers of the ruling government. This breeds harmful attitudes and mentality among the farmers. Xanadu A cooperative exemplified this phenomenon (see *Textbox 6.2* on page 247).

This situation mimics that of Tagarika cooperative where an irrigation equipment sharing cooperative was turned into tractor, implements and irrigation sharing cooperative. Although the scope of implements sharing broadened, it brought many challenges, which overwhelm the benefits. For the case of Xanadu A cooperative, on the other hand, a vegetable or horticultural cooperative was turned into a tractor, implements and irrigation sharing cooperative. In this example, the government's involvement destroyed a cooperative that existed, and that had a substantial ideological focus. As I spoke to the cooperative leader, a weighty sense of regret was detectable in the way they talked. They wished they had not accepted to change the cooperative.

Textbox 6.1 and *Textbox 6.2* is evidence of a continued hubristic government approach to rural development. Dialogue, a genuine one for that matter, would have solved all these issues. They needed to gather enough information about whether such a program could even work on this particular farm. The fact of the matter is that the Zimbabwe agricultural system is not homogenous, and the government does not seem to acknowledge this fact. Blanket policies are always formulated and implemented across the country with little regard to their suitability. An easy way to reduce this problem is to decentralise policy making and implementation so that people who know geospatial differences also take part in policy formulation and implementation.

Textbox 6.1: The story of Tagarika Irrigation Cooperative

Chairman, Security and Secretary of the Tagarika Irrigation Scheme
In the beginning we (about 15 farmers) wanted to form a cooperative that specialized in sharing irrigation equipment in 2003. The cooperative was formed because of problems in accessing water around the farm. Since we had a dam nearby, we had no means of pumping the water to the fields, so we had to come together to buy irrigation equipment since it was too expensive to buy as individuals. The farmers produced various crops, including horticultural crops and maize and soybean. A whose primary function was to lobby the government and other various possible financial sources to get more equipment was established. There was a government program for tractors, but we needed to be registered, and also, we needed to include everyone in this farm to benefit from this program. So, we applied, and the government came with a template for the creation of by-laws, we used that format to create our own by-laws which were checked by the registrar and approved in 2014. We then received two of each of tractors, discs, ploughs, planters, four hosepipe reels and three pumps (to add to the three pumps that we already had).

The cooperative works in this way, farmers pay for using the tractor services, for example, people pay for tillage services of the tractors. Since we got these tractors and other equipment on credit, the money received from the payments by farmers is used to service the machines and pay for consumables, the remainder (profit) is then paid to the Brazilian account. The advantage is that after five years, disregard of amount paid, the tractors become ours (the cooperative). However, this equipment cannot support all the 48 cooperative members. For example, the four hose reels have a 40-hectare irrigation capacity, and given that there are about 200ha for the whole farm, it would be difficult to rely on the four hose reels. It would have been wiser to get plastic pipes which are way cheaper and are better adaptable to single farmers with 3-5 hectares. These hose reels are ideal for a farmer with 40 hectares in which they do not need to move the hose reel too frequently. Thus, the current equipment is inappropriate for the type of farmers that we have and for the scope and scale of production underway.

Old members used to pay monthly subscriptions, and it was easier to follow them up. Now no one pays monthly subscriptions, they only pay when they use the tractors. We had our own irrigation equipment sharing cooperative, but we had to change it to a tractor equipment cooperative to be able to take part in the program. Ideally, we would want a cooperative for irrigation in which the correct type of equipment is acquired, smaller plastic pipes to go with our small-scale production levels instead of centre pivots which were available in the program. There was no dialogue between the farmers and the Brazil tractor program, and we only joined because we wanted tractors, and now, we are trapped. If only the cooperative and the government can be complimentary.

Biggest challenge is that there are low levels of **trust** in the cooperative. However, sometimes it is not the fault of the cooperative leaders; for example, some farmers do not trust us because we could not tell them exactly how much we owe the government and the Brazil account. This is because this information was never given to us as well. In fact, we are still unclear about how the program really works, four years down the line. We were just told that we have to pay for five years, and after this time, it does not matter how much we would have paid, but the tractors will become ours. They did not tell us what the initial amount of money was, or what would happen if we fail to repay the amount in full within the five years. Or what would happen if we overpay? Some of the equipment is not durable and has since broken down.

However, **we do not want to be given equipment for free. We will get used to handouts if that happens, we want to pay, we just want friendly and conducive payment terms. That is all we ask for as farmers of this cooperative** (bold by author for emphasis)

246

Textbox 6.2: The story of Xanadu A Tractor Cooperative

Top leader of the Xanadu A tractor cooperative
Before this cooperative came into existence, it was called Back to Eden and had only eight members. It specialized in fruits and vegetables and had an ideology that encouraged vegetarian diets, hence the name Back to Eden. Our market target was local, and eventually to be able to export our produce. We were actually on track to achieving these goals only two seasons into the cooperative as we had secured markets at Holiday Inn and other like hotels and outlets. Our cooperative had eight members that paid $10 per month as subscription. This translates to $960 per annum just from subscriptions. Everyone produced by themselves with strict supervisory from the cooperative members in order to meet and maintain a set standard acceptable to the markets. The cooperative was an excellent concept.

When the district officers knew about the cooperative, they offered us with implements that were coming from Brazil. However, in order to receive the implement, we had to include everyone in Xanadu farm because this was a government program that was supposed to benefit everyone. About five years ago, the government encouraged us to start the tractor cooperative, and we had to abandon our fruits and vegetable ideology since many of the new people did not understand it, and were not paying their membership fees. We currently have about 103 members, in which more than half are not full members because they are not up to date in their subscription payments. Thus, the introduction of the tractors into the cooperative changed the cooperative for the worse, I think. The merging of committees was not smooth, and some of the incoming members did not trust the existing structures and wanted a complete overhaul of the committees, this caused a lot of problems. Simply put, there were many differences in terms of ideology, trust and focus of the cooperative. Focus was lost in the sense that the current cooperative is not concerned about being productive, but they are concerned about whether the farmers pay for the use of the tractors and do not really care about whether they go on and become productive, as long as the money for servicing is paid. At the moment, farmers are charged $100 per hectare, 60% of this money goes to the Brazilian account, and the rest goes into servicing of the tractors. I think the cooperative movement needs to have almost the same type of people faced with the same type of problems for it to be united and for it to work. Focus should not be on whether one is a landholder or not because most people who hold land are not farmers. Or the fact that it is a government program and hence everyone can join without capacity to produce.

We were not told of how much we had to pay in total and actually we never asked. We just know that after five years; the equipment and tractors will become ours. In terms of clear-cut problems, there are a number of cell phone [absentee] farmers who are not stationed on the farm and are not producing anything, these cause problems. **This means there are many differences between the people. And thus, each farmer's goals and objectives are too dissimilar to the other. Smaller groups or cooperatives are a solution, like-minded people together at first and then eventually they can become bigger** (bold by author for emphasis). AREX [government extension department] should be involved but within smaller groups, and they should focus on extensive training in terms of using implements, tractors, business management.

Summary of chapter

The chapter discussed at great length, the political economy of the cooperative movement in Zimbabwe. The government has too much say in the movement, not only through the Cooperative Societies Act (1996), but they also had a commanding presence in the day-to-day activities of the cooperatives. I discovered that there had been a couple of attempts to revise the cooperative act and give the National Federation of Cooperatives a more central role in the movement. However, these attempts did not see the light of day as the 2005 proposal, and the 2017 proposal was shelved indefinitely. From the various interviews that I carried out, I managed to validate some of the results obtained from the quantitative analysis.

While many things made the cooperative an appealing development model, there exist many other problems that currently hinder the success of the cooperative approach in Zimbabwe. At the top of these is the lack of a stable identity for the cooperative. There is no strong ideological standing which has made it easy for the government and some politicians to infiltrate them through offering free financial benefits. The harsh macro-economic environment has worsened the situation that the country has been experiencing over the last 20 years. It was reported that the movement had much mistrust in the management and leadership committees. Reduced management levels seemed to affect the cooperatives in the CA more.

Farmers were encouraged to change their attitudes about cooperatives, realise that cooperatives do not provide free goods and services. The government should revise the cooperative act, formulate sector-specific acts and give some space to the national cooperative movement to thrive on its own. These cooperatives will be dissimilar to the British-Indian type which are controlled by a hubris government or NGO, but these should take a bottom-up form. Although the new wave has also faced several management challenges, they have the potential to revive the cooperative movement, solve contradictions brought about by the free-markets, and hence resolve the contemporary agrarian questions.

This chapter discussed the farmers' narratives and examined the legislative tool for cooperative; the following chapter will try to interrogate quantitative data in order to understand the position of the cooperative 20 years after the FTLRP. This will help understand the potential for creation and maintenance of robust cooperatives from a more statistical point of view.

Chapter Seven

Peasant Differentiation and Its Effects on Social Economic Production

Introduction

The main objective of this chapter is to discuss and explain the position of the cooperative movement based on quantitative statistical modelling. Before suggesting solutions to its problems, it is necessary to analyse it further. Japan and Zimbabwe were compared from a literature perspective (Chapters Three and Four respectively) and also based on new field evidence (Chapters Five and Six respectively), however, Chapter Seven is a stand-alone chapter without a corresponding Japanese comparative side. While I have presented narratives from farmers, experts, and government officials in Chapter Six, now I ask, "what do the numbers say?" I ask how the class character of the farmers in Goromonzi district affects labour, land, inputs utilisation, income, and production behaviour of farmers. And then examine how these differences affect potential participation in the cooperative. Chapter Seven presence an empirical study of the class differentiation in agrarian villages of Zimbabwe which is critical for the application of the Chayanovian theory of cooperatives. Classification analysis was critical to determine the type of farmers available for the formation of cooperatives as dictated by Chayanov (1991) (see Chapter Two, page 52). The ultimate goal was to enrich my understanding of the current cooperatives as a way to develop a cooperative model in Chapter Eight.

Differentiation within the cooperative movement

Basing on Chayanov's theory of peasant cooperatives, which underpins this book, I sought to understand the socio-economic structure found in the cooperative movement. Socio-economic differentiation has a profound effect on the potential of an agrarian community to form and maintain viable cooperatives. As discussed in Chapter Two, for example, an agrarian structure that is bloated with impoverished peasants who are less productive, with no means of production and who rely on wage labour is unlikely to have successful and sustainable cooperatives. Additionally, a structure that has a substantial proportion of wealthy peasants, who primarily depend on income derived from sources outside of agriculture will have fewer incentives to start cooperatives, and if they do, they will do only to exploit non-rich peasants. In this sense, a clear understanding of the composition of the agrarian structure is essential.

It is imperative for me to explain and prove the robustness of the quantitative analysis process since it is virtually the ground on which the proceeding analysis rests upon. Quantitative data for Zimbabwe was collected from Goromonzi district through a multi-stage sampling which involved purposive snowball sampling of households in the A1 resettled areas as well as in the CA (see Chapter One, page 18). I used data obtained from the 192-respondent questionnaire from Goromonzi district. The quantitative data was used to carry out multivariate analysis (first-factor analysis and then cluster analysis) to group farmers into distinct class categories.

I analysed data from cooperatives from *i)* the resettled areas (A1), i.e., those formed after the land reform era of 2000 and *ii)* the CA mainly formed before the 2000-land reform. Although some argue that A1-peasant is an extension of the CA-peasant, there are significant differences in terms of labour hiring, land sizes, production, and tenure relations. Another reason for separating these in the analysis is that cooperatives in the A1 are relatively newer than those in the CA. I focused on the level of organisation attained by the cooperative, i.e., the robustness of the management committees. The rationale for this is that the commonly identified cooperative problems such as free-rider problems, control problems, adaptation to innovations and technological developments (see Ortmann & King, 2007, pp. 57–59) all depend on the robustness of the cooperative management structures.

252

Factor analysis

Based on Chayanov (1991), I came up with a total of ten variables that directly affect farmers participation in a group organisation (*Table 7.1*)

Table 7.1: Variables used to determine critical study components (N=192)

	Mean	Std. Deviation
Land utilization % (cropped/owned x 100) in 2017	72.91	40.84
Income realised from capital invested outside agriculture (2014-2017)	414.09	2500.98
Average family labour hired in 2014-2017	3.61	1.40
Income from agricultural production in 2014-2017	1816.08	3836.42
Permanent hired labour average of 2014-2017	0.37	1.03
Casual hired labour average of 2014-2017	2.96	4.58
Average amount of family labour sold out of 2014-2017	0.16	0.59
Cropped area in 2017	2.02	2.02
Average savings per household in 2017	236.09	1168.62
What is your meeting % attendance rate?	59.02	19.35

Source: Compiled by the author based on SPSS Factor analysis output, own study

To check the fitness of the data to perform principal factor analysis, I carried out a Kaiser-Meyer-Olkin and Bartlett's test of sampling adequacy in SPSS version 25. The KMO and Bartlett's test of sphericity score of 0.667, and a statistically significant chi-squared test score of p=0.000 showed that the variables chosen were adequate to create groups within the sample.

Using Principal Component Analysis (PCA), we identified four main variables that could explain the differences between the farmers we had spoken to. These became *i)* casual hired labour, *ii)* income from capital invested outside agriculture, *iii)* the average amount of labour sold out, and *iv)* percentage land utilisation rate. These were the results of the factor analysis, which I then employed to carry out the cluster analysis.

Cluster analysis

The four variables identified by the PCA became the inputs of the SPSS Two-step cluster analysis algorithm, which identified five clusters. Because of the quality of the data, the quality of the cluster was well within the acceptable range, as shown by the Silhouette measure of cohesion and separation (*Figure 7.1*). The five clusters contained 4 (2.1%), 15 (7.8%), 31 (16.1%), 66 (34.4%) and 76 (39.6%) farmers, respectively. The most influential variables in determining the clusters were 'amount of family labour hired out' of the household, the amount of 'casual hired labour' used by the households and rate of 'land utilisation'.

Figure 7.1: Two-step cluster model summary

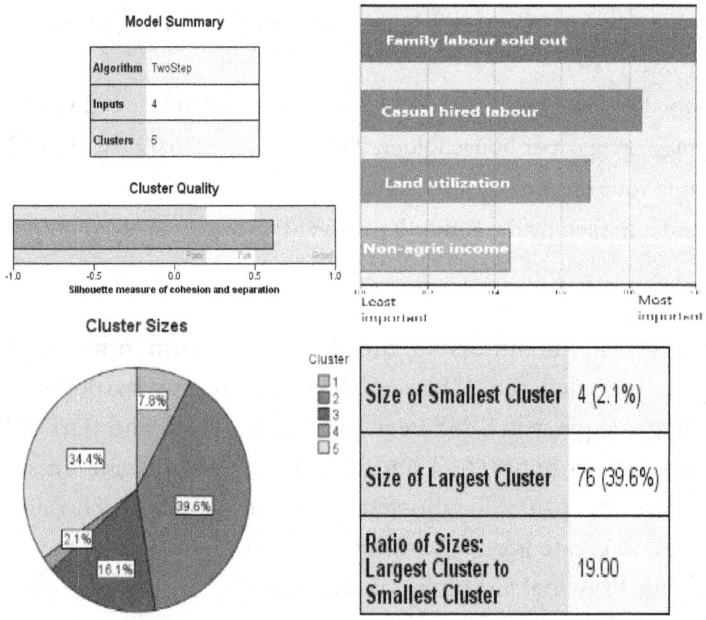

Source: Compiled by the author based on SPSS Cluster analysis output, own study

The least determining variable among the four was the amount of non-agricultural based income' that each household received. The next step was to describe these categories or clusters and come up with suitable names.

In order to understand how farmers differed within the CA and the A1 resettled farming areas, I compared the individual characteristics of the clusters using another output from SPSS (*Figure 7.2*). The first cluster had the highest average amount of family members selling their labour to others within the agrarian structure. These households had low levels of casual labour hiring, income from trade turn-over and land utilisation rates.

Figure 7.2: Cluster composition and characteristic

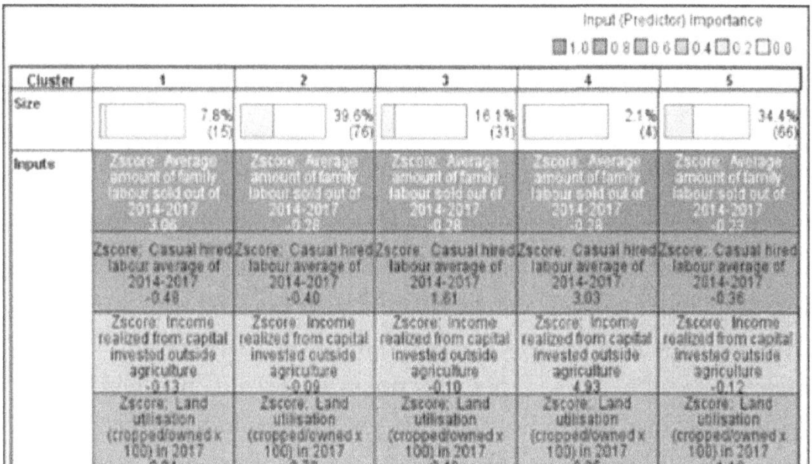

Input (Predictor) importance
1.0 0.8 0.6 0.4 0.2 0.0

Cluster	1	2	3	4	5
Size	7.8% (15)	39.6% (76)	16.1% (31)	2.1% (4)	34.4% (66)
Inputs	Zscore: Average amount of family labour sold out of 2014-2017 3.06	Zscore: Average amount of family labour sold out of 2014-2017 -0.28	Zscore: Average amount of family labour sold out of 2014-2017 -0.28	Zscore: Average amount of family labour sold out of 2014-2017 -0.28	Zscore: Average amount of family labour sold out of 2014-2017 -0.23
	Zscore: Casual hired labour average of 2014-2017 -0.48	Zscore: Casual hired labour average of 2014-2017 -0.40	Zscore: Casual hired labour average of 2014-2017 1.81	Zscore: Casual hired labour average of 2014-2017 3.03	Zscore: Casual hired labour average of 2014-2017 -0.36
	Zscore: Income realized from capital invested outside agriculture -0.13	Zscore: Income realized from capital invested outside agriculture -0.09	Zscore: Income realized from capital invested outside agriculture -0.10	Zscore: Income realized from capital invested outside agriculture 4.93	Zscore: Income realized from capital invested outside agriculture -0.12
	Zscore: Land utilisation (cropped/owned x 100) in 2017 0.04	Zscore: Land utilisation (cropped/owned x 100) in 2017 0.70	Zscore: Land utilisation (cropped/owned x 100) in 2017 0.48	Zscore: Land utilisation (cropped/owned x 100) in 2017 0.86	Zscore: Land utilisation (cropped/owned x 100) in 2017 -1.10

Source: Compiled by the author based on SPSS Cluster analysis output, own study

The farmers in this cluster represent those that hire little labour, are the least productive but also sell their labour to other households and are victims of all types of exploitation. Some of these may not possess the land, and if they do, production occurs within the confines of subsistence production, their income is based solely on selling their labour to other farmers, these usually will not join the cooperative because they lack the minimum resource requirements to participate in the production as highlighted by Chayanov (1991). I named this category of farmers to be the 'penury peasants' (permanent-labourers), meaning indigent farming households.

The second cluster had the lowest average family labour sold and the second-lowest amount of casual labour hired, at the same time. They had little to no alternative sources of income outside of agriculture. However, they had the second-highest land utilisation rates. Thus, these farmers have limited resources for agricultural production, and they remain poor because of this ability to crop all their given land. Thus, the farmers in this cluster exhibit characteristics of those that do hire minimal labour, are productive but do not sell labour. They use their labour exclusively on their farm and get exploited each only when they engage in agricultural commodities markets. This group of farmers are also subject to exploitation from capitalist farmers or commercial companies from the cities through such conduits as contract farming, usury credit lines and loans as explained by Mazwi and Muchetu (2015). I named them 'poor peasants', meaning those that barely manage to feed themselves, *ceteris paribus*, but may need assistance in times of poor weather.

The third cluster consisted of farmers who hardly sold their family labour to other farms but had high rates of casual labour hiring. Their incomes and land utilisation rates were intermediate. The farmers in this cluster represent those that have diversified income streams. These types of farmers are usually involved in petty trading and may be employed in the semi-skilled industry while at the same time undertaking commercial or market-oriented agricultural production. They hire labour, usually on a part-time basis. These are at more risk of exploitation from the market than from the capitalist farmers. I named them 'middle-rich peasants' in this study.

The fourth cluster of farmers had low levels of family labour sold out to other farms. However, they had the highest average amount of casual labour hiring (paid labour), the highest amount of income from investments outside agriculture and the highest rates of land utilisation. These farmers are the prosperous farmers. These types of farmers usually hire-in agricultural land, which resulted in their land utilisation rates exceeding a hundred per cent. They usually have a business in the growth points and maybe running transportation, maize milling, and related services in the rural areas. They obtain income from trade turnover (returns on investment outside agricultural production), sometimes avoid hiring farm labour for agricultural production on their own farms (they may hire labour for non-farm activities). They are influential in the community, countryside and hence in the cooperative. Their numbers are, however, limited in Southern African agrarian structures, as evidenced in this study; they represent only 2.1% of the sample. I called this group of farmers 'rich peasants' (capitalist peasant).

The fifth and final group of farmers was composed of households that scored lowest in land utilisation as well as selling of own-family labour. Their levels of casual labour hiring and income from investments outside of agriculture were also low. These farmers solely depend on agricultural production using limited resources and are a target of exploitation from all agents of exploitation. They form the bulk of the farmers in this study and also in African agrarian structures (Amin, 2012). I referred to them as the 'middle peasants' (efficient land users) in this book. The essential characterisation criteria were the average amount of family labour sold for cluster one, casual labour hired for cluster two and three, casual labour hired and income from investment for cluster four, and land utilisation for cluster five.

These different classes were named based on a modification of Chayanov (1991), which had six categories (see page 60). In the Chayanovian categorisation, the sixth type of peasant had no access to land and was not involved in agricultural production, hence survived only by selling their labour. I did not have these types of farmers in my survey, especially in the newly resettled farming area where land ownership was a prerequisite. Thus, each name or category was informed by the composition or intensity (SPSS score) of the four variables, as explained in this section.

After completing cluster analysis, I carried out a Pearson Chi-squared test and a comparison of column proportion analysis of the clusters for the settlement or model types. A higher proportion of farmers in the A1 sector are middle peasants (45.7%) followed by the middle-rich peasants (31.5%) while rich peasants only accounted for 3.3%. Penury peasants were few in both models. The CAs cooperative movement has more poor peasants (64%) followed by middle peasants (24%) while rich peasants were hardly present (see *Figure 7.3*). The results of my classification model revealed three dominating (and two passive) classes.

Figure 7.3: Custom pivot tables, Chi-square test and comparison of proportions analysis

	Settlement type					
	A1		CA		Total	
Peasant class	No.	%	No.	%	No.	%
Penury	6	6.5	9	9.0	15	7.8
Poor	12	13.0	64	64.0	76	39.6
Middle	42	45.7	24	24.0	66	34.4
Middle-Rich	29	31.5	2	2.0	31	16.1
Rich	3	3.3	1	1.0	4	2.1
Total	92	100	100	100	192	100

Pearson Chi-Square Tests

		a.9. Model type
Classification of the peasants	Chi-square	65.384
	df	4
	Sig.	.000[a,b]

Comparisons of Column Proportions[a]

		a.9. Model type	
		A1	CA
		(A)	(B)
Classification of the peasants	Penury		
	Poor		A (.000)
	Middle	B (.002)	
	Middle-Rich	B (.000)	
	Rich		

Source: Compiled by the author based on SPSS Pivot table analysis output, own study

The composition of the clusters within the two settlement models were statistically different, as highlighted in the Pearson chi-squared test (it was significant at a 95% confidence interval – CI). I found statistically significant differences between the poor, middle and middle-rich clusters with significance values of 0.00, 0.02 and 0.00 at 5% CI respectively. The penury and rich peasant clusters were not significantly different between farmers in the A1 and those in the CAs. This finding seems to confirm the existence of three significantly different dominant classes within the peasantry pointing to how the national trimodal agrarian structure (Moyo et al., 2009) could establish itself at local levels (see also Mazwi et al., 2021 upcoming).

Socio-economic production trends and patterns in peasant movements

The theory of peasant cooperatives emphasised that the success of cooperatives depends on developing linkages between farming organisations (Chayanov, 1991, p. 225). The emphasis is on the presence of organisational structures that enable the implementation of socio-economic ideas within the cooperative. The unit of analysis in this subsection is the cooperative itself.

Socio-economic status of the cooperators.

The average membership of the nine cooperatives was 71 and was slightly higher in the CA (75) than in the A1 sector (66). Female membership outnumbered males 57.7% to 42.3%. However, there was a statistically significant difference between females in A1 (37.9%), and the CAs as more females (70.7%) were members of the cooperative groups (*Table 7.2*).

Table 7.2: Gender and marital status of cooperative members

Variable	A1		CA		Total	
Gender	mean	%	mean	%	mean	%
Male	41	62.1	22	29.3	30	42.3
Female	25	37.9	53	70.7	41	57.7
Average total	66	100	75	100	71	100
Marital status	**No.**	**%**	**No.**	**%**	**No.**	**%**
Married	68	73.9	67	67	135	70.3
Single	1	1.1	0	0	1	0.5
Divorced	1	1.1	3	3	4	2.1
Separated	1	1.1	6	6	7	3.6
Widowed	21	22.8	24	24	45	23.4

Source: Compiled by the author based on own survey data, 2018-19

Many of the cooperative members (70%) were married across the two settlement models, followed by those who were widowed (23.4%). Although 82% of the widowed cooperative members were females, there were no significant statistical differences in marital status between the two models. I discuss the implications of lower women participation later (page 304).

Given the proximity of the district to the capital city of Harare, some studies in the same area found that farmers straddle between farming and non-farm wage labour in the city (Chambati, 2017). However, approximately 83% of the cooperators were unemployed and hence were full-time farmers earning their livelihoods from agricultural and on-farm production (*Table 7.3*). Furthermore, 17.7% of the respondents had never earned any income from formal sources and have thus always relied on farming. Of the 65.6% formerly employed, only 11.3% were receiving pensions. This information is particularly important in defining the farmer or peasant. Additionally, the peasant cooperative theory appeals more to peasant societies with limited livelihood options (Chayanov, 1991).

Table 7.3: Employment status of members

		A1		CA		Total	
Employment status		No.	%	No.	%	No.	%
	Never been employed	22	23.9	12	12	34	17.7
Unemployed	Formerly employed	59	64.1	67	67	126	65.6
	Total unemployed	81	88	79	79	160	83.3
Currently employed		11	12	21	21	32	16.7
Receiving pension		8	9.9	10	12.7	18	11.3

Source: Compiled by the author based on own survey data, 2018-19

The level of education and technical skills attained through agricultural technical training is vital in management or assimilation of information. The majority of the co-operators (96.4%) were literate, meaning that they could at least read and write and had managed to complete at least seven years of primary education (*Figure 7.4*). Only 3.6% of the respondents had no formal education. There were no statistically significant differences between the two settlement models.

Figure 7.4: Level of education attained, and formal agricultural training received.

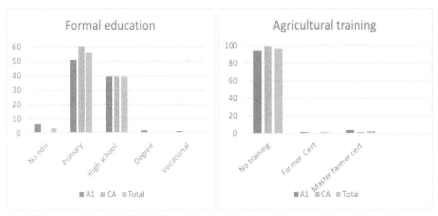

Source: Compiled by the author based on own survey data, 2018-19

Very few of the respondents (3%) had received formal agricultural training (with a certification document). In-depth interviews revealed that cooperatives carry out member training on the production of various agricultural products regularly but did not give certificates except on rare occasions where certificates of attendance are issued. My data, therefore, illustrates that the level and quality of education and training were sub-optimal. This, in turn, affected flow and comprehension of information, hence affected the quality of management committee members.

Establishment of cooperatives

So, what happens when cooperatives are being formed in rural areas? What are the differences between the two settlement areas? Cooperatives are primarily determined to improve or develop the lives of the members. Overall, to improve the standard of living (33.3%); to empower marginalised members (21.9%), to address marketing imperfections (15.3%) and to increase productive farmer capacity (14.4%) were the primary reasons for establishing cooperatives (*Figure 7.5*). There were no statistically significant differences between the CA and the A1.

Figure 7.5: Reason for forming cooperatives.

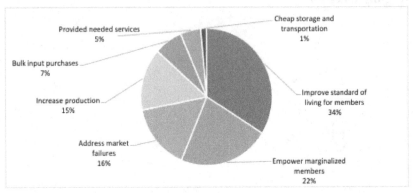

Source: Compiled by the author based on own survey data, 2018-19

Comparison between these findings and the reasons why individual members joined the cooperative reveals a slight disconnection between reasons why members join cooperatives and reasons for the establishment of the cooperative in the first place. The majority of the people joined in order to increase their level of income or their production (56.3%), followed by those who joined in order to improve the quality of their product (24%), increasing the bargaining power (6.8%) and as a defence mechanism against adverse market conditions (5.7%) were ranked third and fourth respectively (*Table 7.4*). This disjuncture between the goals of the cooperatives and those of the individual members is a potential source of inefficiency and may indeed lead to what Ortmann & King (2007) identified as control problems between the principals and the membership.

Additionally, most A1 cooperators had joined the cooperative in order to improve the quality of their produce (47.8%) while those in the CA wanted to increase the quantity of their produce (75%). Thus, the reason for establishing cooperatives seems generic, but the motives that eventually drive farmers into joining the cooperative are diverse. That may indicate the reason why multi-purpose cooperatives are more attractive to farmers than single-purpose cooperatives. This data has implications on the development of strategies to improve the current cooperative movement as well as on the cooperative model I intend to produce.

Table 7.4: The main reasons for joining Cooperative

Reason	A1 No.	%	CA No.	%	Total No.	%
Increase production	33	35.9	75	75	108	56.3
Improve product or service quality	44	47.8	2	2	46	24
Increase bargaining power	0	0	13	13	13	6.8
Defence against adverse conditions	10	10.9	1	1	11	5.7
Lower operating costs	1	1.1	5	5	6	3.1
Bulk purchases	0	0	4	4	4	2.1
Obtain services otherwise unavailable	3	3.3	0	0	3	1.6
Cheap storage and transportation	1	1.1	0	0	1	0.5
Total	92	100	100	100	192	100

Source: Compiled by the author based on own survey data, 2018-19

Historically, cooperatives in Africa were formed either by the colonial governments or later in the 1990s by NGOs and other development agents. In the CA, 64% of the respondents belonged to cooperatives that were formed by NGOs or donor organisation then followed the government (20%).

Table 7.5: Founder of the cooperatives

	A1 No.	%	CA No.	%	Total No.	%
Political Party	1	1.1	0	0	1	0.5
Extension Officer	0	0	1	1	1	0.5
NGO	1	1.1	64	64	65	33.9
Local Authority	0	0	4	4	4	2.1
Local Political Leader	0	0	4	4	4	2.1
Local Farmers or Coop Members	90	97.8	7	7	97	50.5
DDP (Government)	0	0	20	20	20	10.4
Total	92	100	100	100	192	100
Chi-square= 162.030 df = 6 sig.= 0.000*						

263

Source: Compiled by the author based on own survey data, 2018-19

The results presented in Table 7.5 are in stark contrast to cooperatives in the A1 model in which 97.8% of the respondents reported that their cooperative was formed and driven by the local farmers or the cooperative members themselves.

I found statistically significant differences in what cooperative members perceived as factors that affect a member's ability or chances to join a cooperative. In the CAs, socio-economic status (reported by 56.6% of the cooperators), level of production (56%) and land ownership (42%) were the three significant factors that could affect a member's interest in joining and of being admitted into the cooperative (*Table 7.6*). It is fascinating to find out that the majority of CA participants listed socioeconomic status (peasant classification) as the significant factor affecting admittance into the cooperative. The peasant class that an individual placed themselves in affects their decision to join or form a cooperative. It also affects the type of people that can join a cooperative. This fact strengthens the rationale for carrying out a robust social differentiation analysis as done on page 251.

Table 7.6: Determinants of someone joining a cooperative?

	A1		CA		Total		Chi-square test		
	No.	%	No.	%	No.	%	Sig.	Df	X² value
Farm/Land ownership	91	98.9	42	42	133	69.3	.000	1	72.914
Level of production	2	2.2	56	56	58	30.2	.000	1	65.848
Socio-economic status	1	1.1	56	56.6	57	29.8	.000	1	70.102
Political affiliation	0	0	14	14.1	14	7.3	.000	1	14.039
Level of education	1	1.1	28	28.3	29	15.2	.000	1	27.387

Source: Multiple responses; Compiled by the author based on own survey data, 2018-19

Divergent views emerged from most A1 cooperators who highlighted that only land/farm ownership status (98.9%) might affect chances to join or to be admitted into a cooperative. Although a few CA farmers reported that political affiliation (14.1%) and level of education (28.3%) affected chances of joining a cooperative, less than 3% reported these two reasons in the A1 sector. Although there is a need for more research in this aspect, what is evident from these results is the fact that there exist different hurdles for joining a cooperative between the CA and the resettled areas.

Additionally, I examined whether the members had indeed enjoyed any benefit ever since joining the cooperative. A thought-provoking result across the two settlement types is that approximately one-fifth of the cooperators highlighted that they had seen no difference (*Figure 7.6*). The difference between the reasons for forming the cooperative and the reason for joining that cooperative, as discussed earlier, might explain why some did not find the cooperative helpful. Cooperatives need to carry out more training to conscientize their members on cooperative goals and objectives.

Figure 7.6: Benefits experienced after joining cooperative.

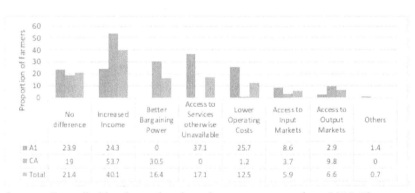

Source: Compiled by the author based on own survey data, 2018-19

However, of the 79.6% of the respondents that actually experienced a substantial improvement after joining the cooperative, increased income (40.1%), better bargaining power (16.4%), access to services otherwise unavailable (17.4%) and lower operating costs (12.5%) were mentioned across both farming models (*Figure 7.6*). This is an important finding since we argued that cooperatives could resolve the AQ. Not only were cooperatives able to improve income and bargaining power, but they were able to provide new services that were not available, proving the ability of the cooperative to

265

innovate. The CA movement scored higher in terms of improving income and consolidating the bargaining power of the peasants, which is ideal for fighting the adverse effects of the free market system. The A1 farmers, on the other hand, also enjoyed increased income (24.3%) and lowered operating costs (25.7%). Nonetheless, most had benefited from the ability to access services that were otherwise unavailable in the area (37.1%). In this case, cooperatives are filling in a gap that the free market, as well as government programs, have failed to do and thus validates Hayami Yujiro's (2005; 2010) CMS framework.

The flow of information in the cooperative

The proportion of farmers who knew such things as frequency of meetings, awareness of the year of cooperative establishment, the number of members (by gender) in the cooperative as well as sources of funding averaged over 80% across the two settlement models. Thus, members accessed a considerable amount of necessary information. However, such information as the cooperative objectives, mission, and goals, which is critical in uniting the people towards achieving set goals was not widely accessible to the rank-and-file membership. Just over 56% of the respondents received this type of information through training, implying that almost half of the cooperative members did not formally receive and were virtually unaware of this mission, objectives, and philosophy information (*Table 7.7*). Considering that A1 cooperatives are relatively new, hardly ten years since formation, the fact that the proportion of members who accessed information is approximately equal to that of CA is commendable.

Except for the amount of cooperative debt (69.3%), access to financial information was relatively low as seen through those who reported having knowledge of the amount of money paid to apex organisations (38%), annual balances of the reserve ratio (28.6%) and how much each share in their cooperative cost (3.1%) (*Table 7.7*). Therefore, it appears that general information such as the number of people in the cooperative is more natural to get, but finance related information is harder to get within both CA and A1 cooperatives. Lack of transparency leads to much mistrust within the cooperative

266

movement since sensitive information on the financial health of the cooperative should be known to every cooperative member. This problem was also found in the Japanese Sanbu network case study.

Table 7.7: Access to information within the cooperative

Type of information	A1 No.	%	CA No.	%	Total No.	%
Frequency of meetings	87	94.6	100	100	187	97.4
Year of establishment	89	96.3	97	97	186	96.7
Membership structures (by gender)	81	88	100	100	181	94.3
Sources of funding	86	93.5	76	76	162	84.4
Amount of cooperative debt	44	47.8	89	89	133	69.3
Goals and objectives	51	55.4	58	58	109	56.8
Total money paid to Apex organisations	11	12	62	62	73	38
Annual balances in the revenue fund	2	2.2	53	53	55	28.6
Price of shares	0	0	6	6	6	3.1

Source: Compiled by the author based on own survey data, 2018-19

Cooperative management issues

The Cooperative Societies Act of 1996 has specific provisions that stipulate that all cooperatives should have a management committee. The efficiency of this committee depends on several variables such as level of education; skills; internal and external flow, interpretation, and assimilation of information as well as the levels of trust. A few critics of the cooperative movement cite, relative to corporate professionals and well-trained managers, the inefficiency of the management committee as a rate-determining-factor in cooperative success (M. Cook et al., 2009; Ortmann & King, 2007). The management committee is one of the mandatory structures that should exist in any cooperative. As I highlighted in Chapter One and Two, gender participation in the cooperative is also of great importance in understanding the potential of the cooperative to answer the agrarian question of gender (and hence the overall agrarian question). Overall, for every five female members in a management committee, there were six male members (or 45.5% of the members in the committee). In the CA, the ratio was higher (5:4 or 59%), as compared to the A1 (5:9 or 30.8%) (*Figure 7.7*).

267

The relatively lower number of women in the A1 cooperative can be explained by the fact that acceptance into the cooperative significantly depended on land ownership, and also that women advocacy groups had campaigned for a minimum 20% land allocation to women (19.5% as reported in SMAIAS (2015) and 24% in Utete (2003)). Thus, more women are taking part in the cooperative movement than they would in a government or private-sector program suggesting that women capacitation would be higher within a cooperatives set-up. This result illustrates how the cooperative model can help in solving some of the gender issues in agriculture, such as women access to land and production resources. Women participation in the cooperatives is particularly encouraged by the ability of the cooperative to give a platform for the women to express themselves (giving them a voice, respect, political legitimacy and influence) (Green, 2013). In the context of agriculture, therefore, women access to input/output marketing and production services is significantly improved through the cooperative. Thus, cooperatives enable isolated women issues to be mainstreamed. Additionally, women-only cooperatives are ideally structured to provide a free and comfortable space for women to lead and develop themselves (ILO, 2015). These are among the primary motivations for women to join the cooperative.

I also collected opinions of cooperative members in terms of how they felt about women integration within the cooperative structures and the cooperative as a whole. Approximately 63.6% reported that the cooperative had integrated women (integrated and well-integrated), 31.7% reported that there was no significant difference and only 3.7% of them argued that women were not well integrated into the cooperative.

After conducting further analysis, it appeared as though co-operators had low confidence in the management committees in as far as essential aspects of the management process such as professionalism, good governance, dynamism, and mutual trust were concerned (*Figure 7.8*). Interestingly, members felt that their management committees scored well in terms of commitment, non-political affiliation, self-reliance, and energy. Thus, I argue that incompetence explains the low confidence of the membership in the committees in terms of professionalism and good governance. Indeed, such characteristics originate from the lack of agricultural training (see discussion on page 266).

Figure 7.7: Perception of women integration and composition in cooperative committees

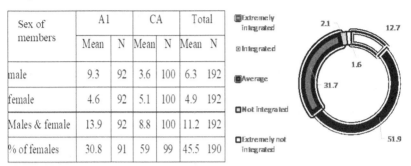

Sex of members	A1		CA		Total	
	Mean	N	Mean	N	Mean	N
male	9.3	92	3.6	100	6.3	192
female	4.6	92	5.1	100	4.9	192
Males & female	13.9	92	8.8	100	11.2	192
% of females	30.8	91	59	99	45.5	190

Source: Compiled by the author based on own survey data, 2018-19

The frequency of cooperative meetings, together with the level of participation of the cooperative members is indicative of the health of the management committee. The more the number of meetings, the more the management can report and spread information to its members as well as to get feedback. Meetings are a crucial way of minimising the free-rider and control problems. Overall, an average of ten meetings per year was held. More meetings were recorded in the CA (13/annum or monthly) than in the A1 model (7/annum or bi-monthly).

Further comparative research is required to determine how these rates compare with those in the rest of Africa and the world. As expected, more cooperators in the A1 (92.4%) than in the CA (25%) agreed that the number of meetings held was not enough, and more meetings were necessary. The quality of resolutions resulting from the meetings depends on the structure and composition of the people who attend the meeting.

Although my study could not get data on whether cooperatives were following strict quorum principles when conducting meetings, I was able to establish that each member attended, on average, 59% of these meetings. Thus, using an average of ten meetings per annum, members were able to attend six of them, or in other words, if the cooperative by-laws stipulate that meetings would be held every month, each member will only attend every other meeting. This has severe implications for the success of the cooperative, especially the newly formed ones in which members would attend, on average, four meetings per year. Thus, managers need to work extra hard in sending information on time and announcing the schedule of the meetings to improve attendance rates.

Figure 7.8: Management committee appraisal by cooperative members

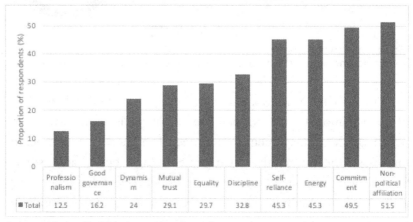

Source: Compiled by the author based on own survey data, 2018-19

Ishida (2003, p. 42) highlighted the existence of the mutual trust and mutual inspection in the management echelons of the Japanese cooperative movement and how suspicion was rooted in the withholding of information, especially in the finance department. For Zimbabwe, approximately 47.1% of the people interviewed said that they were aware of at least one case of corruption that had occurred in the cooperative. Sadly, more cases, though the difference was not statistically significant, were reported in the new established A1 (50.5%) cooperatives than in the CA (44%). FGD data pointed to some members of the cooperatives using political lines to justify their looting of cooperative resources (fertiliser, in that specific case).

At this moment, I learned that the overall political rot at the national level had found its way into household levels as those related to the then ruling echelons abused their power and privilege, and amassed wealth and resources from the people. FGD participants were encouraged to speak out if they saw any corrupt activities by the chairman of an apex organisation who stressed that political demi-gods should be destroyed and that even chiefs can be held accountable. Results revealed that corruption cases were resolved within cooperative structures in the A1 sector (60.9%) while no action was taken for the remainder of the cases (39.1%) (*Table 7.8*). Interestingly, cooperatives in the CA took most of the corruption cases to the police (75%).

Table 7.8: Management's handling of corruption cases

		A1		CA		Total	
		No.	%	No.	%	No.	%
No corruption cases		45	49.5	56	56	101	52.9
Corruption cases	Reported to police	0	0	33	75	33	36.7
	Within Coop structure	28	60.9	8	18.2	36	40
	Nothing	18	39.1	3	6.8	21	23.3
	Total corrupt cases	46	50.5	44	44	90	47.1
Total		92	100	100	100	191	100

Source: Compiled by the author based on own survey data, 2018-19

The need to settle inter-cooperative contradictions internally can be viewed or justified as a necessary evil in order to preserve peace and conflict within both the cooperative and the community. However, failure to effectively recover debts from members is detrimental to cooperatives. From the data obtained from FGDs, this was indeed the biggest challenge that the cooperative movement was facing. Additionally, the cooperative, as a social organisation whose clients are also members, had limited punitive debt collection measures in case of default.

When holding differences in individual cooperative by-laws constant, bad debts were rarely reported to the police (3.1%), and instead, nothing (60.9%) was done about the issue (see *Table 7.9*). The cooperative society hoped the members would, on their own accord, eventually payback. Sometimes the debts were wholly written off as bad debts (18.8%) while a few cases had resulted in the confiscation of debtor's assets (17.2%). There were statistically significant differences between the old and the new cooperative movements, with the former mainly doing nothing (52%). Approximately 70% of the A1 farmers reported that nothing is done in terms of trying to recover unpaid cooperative debts or in the worst cases they are written off as bad debts (28.3%). Therefore, the A1 cooperative movement potentially loses 100% of its bad debts which is hugely detrimental to the sustainability of the cooperative. Better debt recovery mechanisms are necessary in this regard.

Table 7.9: Cooperative debt recovery mechanisms

	A1		CA		Total	
	No.	%	No.	%	No.	%
Report to police	1	1.1	5	5	6	3.1
Nothing	65	70.7	52	52	117	60.9
Confiscate assets	0	0	33	33	33	17.2
Write them off as bad debts	26	28.3	10	10	36	18.8
Total	92	100	100	100	192	100

Source: Compiled by the author based on own survey data, 2018-19

What was not covered in the study was the contents of the individual cooperative by-laws that, as provided by the Cooperative Societies Act (1996), is supposed to have specifications on the prosecution of members suspected of misconduct or corruption. However, to what level do cooperative members have confidence in their management committee in terms of upholding the law? The data obtained suggested that significant numbers of cooperators, especially in the A1 (82.6%) were confident that their management committee had a practical knowledge of the Cooperative Societies law which guides their activities. For the CA, approximately 43% reported that they were sure that the committees understood the act/law. In the same instance, many of the cooperative members believed that their committees were either neutral (41.1%) or did not take seriously (33.9%) the auditing of cooperative accounting books at the end of the financial year (*Table 7.10*). This result is not surprising, given the fact that just over half (55.7%, see *Figure 7.4*, page 261) of the members of the cooperative had attained formal education up to primary level, and the fact that they virtually had received no agricultural training. However, it shows that there is massive potential for the movement if more educational programs or support are infused into their activities.

Table 7.10: Cooperative management and auditing of accounting books

		A1		CA		Total	
		No.	%	No.	%	No.	%
Awareness of Cooperative law	Yes	76	82.6	43	43	119	62
Auditing of accounting books were taken seriously by management'	Larger extent	39	42.4	9	9	48	25
	Neutral	8	8.7	71	71	79	41.1
	Lesser extent	45	48.9	20	20	65	33.9
	Total	92	100	100	100	192	100

Source: Compiled by the author based on own survey data, 2018-19

Cooperators listed several issues that they thought were mostly affecting their management committees. The problems identified by those in the CA differed significantly from those noted A1 landholders. Three major issues in the A1 sector were power relations (52.2%), low mutual trust (17.4%) and an inadequate number of meetings (14.1%). That meant that the management committee had many characters that wanted authority, yet they did not trust each other, and the fact that they did not meet often enough to discuss these issues exacerbated the problems (*Table 7.11*).

The problems identified in the CA were more varied, ranging from quorum issues (28%), low mutual trust (16%), methods of committee member selection (14%) and low levels of skills and qualification (13%).

Given the above discussion, it is not surprising that the A1 sector (30.4%) (with 'power relations' issues) had fewer respondents that

273

thought cooperatives should be headed and run by outsiders who are trained and academically competent than those in the CAs. In contrast, the more experienced co-operators in the CA (56%) seemed to warm up to the idea of incorporating more skilled and professionals into the cooperative structures. This may be indicative of the fact that CA cooperatives, mostly founded by professionals from NGOs and government, are more open to the idea of depending more on external resources than the A1s.

Table 7.11: Concerns raised by co-operators with regards to the efficiency of managers

	A1		CA		Total	
	No.	%	No.	%	No.	%
Quorum issues	2	2.2	28	28	30	15.6
Low qualifications	6	6.5	13	13	19	9.9
Method of selection	6	6.5	14	14	20	10.4
Corruption	1	1.1	9	9	10	5.2
Power relations	48	52.2	1	1	49	25.5
Incompetence	0	0	9	9	9	4.7
Low mutual trust	16	17.4	16	16	32	16.7
Inadequate meetings	13	14.1	10	10	23	12
Total	92	100	100	100	192	100

Source: Compiled by the author based on own survey data, 2018-19

More central to the theoretical debate presented in earlier (Chapter Two) are issues of the role or the relationships forged between players in the agricultural and non-agricultural sectors of the economy of Zimbabwe. I sought to understand the strength of these relations between cooperatives and other sectors such as corporates and the government. The critical thing to learn from *Table 7.12* is the fact that A1 cooperatives had good relations with government institutions (67.4%) and corporate/private companies (41.3%) while co-operators in the CA reported that they had the most substantial ties with the NGOs (94%) and government institutions (83%).

Table 7.12: The strength of relations between cooperative and other sub-sectors

	A1		CA		Total	
	No	%	No	%	No	%
Other agriculture cooperatives	5	5.4	21	21.0	26	13.5
Other general coops	8	8.7	21	21.0	29	15.1
Corporates or private companies	38	41.3	35	35.0	73	38.0
Government institutions	62	67.4	83	83.0	145	75.5
International donors or NGOs	22	23.9	94	94.0	116	60.4
Other non-member farmers	13	14.1	12	12.0	25	13.0

Source: Compiled by the author based on own survey data, 2018-19

This finding has severe implications on the argument advanced in this book that cooperatives can bridge the distance between the state, the private sector and themselves within the CMS framework. Thus, the old cooperative movement, formed basically by the government or by development agents seems to neglect the role of the private sector while the new cooperative movement witnessed in the newly resettled areas is connected to the state and the markets as well as with the multi-lateral institutions (NGOs and donors). This result illustrates the need to reduce binary or polarised models of development.

Effects of class differentiation on cooperative development

In the theory of peasant cooperatives, Chayanov highlights how the problem of differentiation in the peasantry can affect the ability or potential of farmers to join, strive and remain within a cooperative (see Chapter Two). In this sub-section, using empirical data collected from the field, I apply this argument to construct a more vibrant picture of the extent to which the different classes (from my cluster analysis on page 254) participate, engage, or access agricultural market information and market facilities. I divided this section into subsections that reflect the different spheres of the contemporary AQ. My unit of analysis in this sub-section is the farmer or cooperative member.

Access to market information

As noted in Hayami (2005), information asymmetries account for a substantial part of agrarian problems because access to all types of agricultural information is fundamental for the peasants to access cheaper input or output markets and increase their incomes (hence avoid exploitation). Overall, 76% of the respondents reported that they had access to agricultural information and as expected, higher proportions of farmers who accessed information were in the middle, middle-rich and rich peasants. While market price and demand schedules were accessible to the farmers, other types of essential information such as availability of services (seed, insecticide, water, fertiliser and equipment information was harder to find.

Word of mouth from other farmers within the same cooperative (32.5%), Radio/TV (26.3%) and family/friends (18.8%) were the most common channel for all clusters (*Table 7.13*). Farmers also accessed information through SMS messages (5.2%). Radio/TV programs played an important role in terms of disseminating the types of commodities on-demand in different markets, prices of these commodities and the estimated times in which these commodities could be on-demand. An under-resourced farmer without Radio/TV or electric power cannot access this channel, and hence word of mouth is an essential source for them. What this implies is that any information passed through cooperative structures would get more audience than that passed through television. Additionally, the use of the internet is depressed due to low connectedness in rural areas. Radios/TVs, internet and SMSs/cell-phone channels are high potential information dissemination channels that are still grossly underutilised at the moment.

It is not surprising to find out that most of the people in the rich (75%) and middle-rich (67.7%) were happy with the quality of the information that they had access. This is because they paid a premium for the information (TV, radio, cellphones and electricity). On the other hand, many of the poor households in the penury and poor category rated the quality of the information as fair or poor.

Table 7.13: Sources of price information for the cooperative members

	Penury		Poor		Middle		Middle-Rich		Rich		Overall	
	No.	%	No.	%	No.	%	No.	%	No.	%	No.	%
My Coop structures	0	0	6	6.7	3	2.8	3	3.7	0	0	12	3.9
Family and friends	5	21.7	24	27	26	24.5	3	3.7	0	0	58	18.8
Radio/TV	8	34.8	25	28.1	16	15.1	29	35.4	3	37.5	81	26.3
Other Coop Farmers	7	30.4	19	21.3	43	40.6	28	34.1	3	37.5	100	32.5
Market Posters	1	4.3	0	0	0	0	0	0	0	0	1	0.3
Agricultural traders	0	0	0	0	9	8.5	8	9.8	0	0	17	5.5
SMS messages	2	8.7	10	11.2	4	3.8	0	0	0	0	16	5.2
Internet	0	0	3	3.4	0	0	0	0	0	0	3	1
Extension officer	0	0	2	2.2	5	4.7	11	13.4	2	25	20	6.5

Source: Compiled by the author based on own survey data, 2018-19

There has been a long-standing debate within the government on their role in the provision of extension services in the rural areas of Zimbabwe in terms of increasing/reducing its budget allocations while at the same time allowing private sector participation in extension (Hanyani-Mlambo, 2000, pp. 666-668). The fact of the matter is the budget allocation to the sector have always been inadequate, from the time of independence to the time of the disastrous collectives and the post-FTLRP era. Akwabi-Ameyaw (1997) cited poor extension services as one of the main reasons why the 1980s collectives had failed. In recent years, in light of increased agricultural market liberalisation, more state policymakers have considered reducing extension services, presumably, to allow for individual farmers to purchase own extension services from the open market (Hanyani-Mlambo, 2000, p. 665). Although there were no statistically significant differences between clusters, my data revealed that in addition to private extension services, higher proportions of farmers in the upper echelons of the peasant classes accessed public extension services such as the AREX and Research & Irrigation Development.

Table 7.14: Access to government public extension services.

	Classification of the peasants											
	Penury		Poor		Middle		Middle-Rich		Rich		Total	
	No.	%	No.	%	No.	%	No.	%	No.	%	No.	%
Dept of crop extension	10	71.7	61	84.7	49	79	31	100	4	100	158	85.9
Dept of livestock & veterinary	9	60	39	54.2	43	69.4	29	93.5	4	100	124	67.4
Dept irrigation & tech services	10	71.7	59	75.7	60	9.23	28	90.3	3	75	158	83.3
Dept natural resources	0	0	2	2.8	0	0	0	0	0	0	2	1.1
Forestry commission	3	20	9	12.5	1	1.6	0	0	1	25	14	7.6
Private company	6	40	20	27.8	19	30.6	11	35.5	2	50	58	31.5
NGO	7	46.7	41	56.9	21	32.3	1	3.2	1	25	71	38
Farmer Cooperative	0	0	6	8.3	15	23.1	6	19.4	0	0	27	14.4
Local opinion leaders	6	40	18	25	38	58.5	24	77.4	4	100	90	48.1

Source: Compiled by the author based on own survey data, 2018-19

A different scenario characterised the poor and penury farmers as the provision of extension services was dominated by NGOs (*Table 7.14*). With the global push to withdraw the state from agriculture, it would not be surprising to see decreasing government services to the middle, poor and penury classes and an increase in the NGO and private extension activities. In such a case, cooperatives become necessary.

Extension service delivery falls into many categories ranging from farmer training, group development, commodity interest groups, demonstrations, competitions, field days, study tours and farm visits. These can be regrouped into; i) the group approach, ii) individual approach (in which each extension officer visits each farmer); iii) and use of other channels which may include mass media (radio & television) (Hanyani-Mlambo, 2000, p. 667).

Extension officers hardly utilise one approach, frequently using a mixture of the three depending on the type of information to be disseminated. I found significant differences between extension approaches used between the five peasant classes.

Table 7.15: Use and frequency of public extension services by cooperative members

	Penury		Poor		Middle		Middle-Rich		Rich		Total	
	No.	%	No.	%	No.	%	No.	%	No.	%	No.	%
Number of times group approach was used in extension												
0	1	6.7	25	32.9	15	22.7	0	.0	0	.0	41	21.4
>2	8	53.3	15	19.7	15	22.7	7	22.6	1	25.0	46	24.0
2-6	6	40.0	33	43.4	36	54.5	24	77.4	2	50.0	101	52.6
<6	0	.0	3	3.9	0	.0	0	.0	1	25.0	4	2.1
Number of times individual approach was used in extension												
0	5	33.3	28	36.8	27	40.9	23	74.2	3	75.0	86	44.8
>2	7	46.7	45	59.2	35	53.0	7	22.6	0	.0	94	49.0
2-6	3	20.0	3	3.9	4	6.1	1	3.2	1	25.0	12	6.3
Number of times mass media approach was used in extension												
0	15	100.0	75	98.7	64	97.0	31	100.0	4	100.0	189	98.4
>2	0	.0	1	1.3	2	3.0	0	.0	0	.0	3	1.6

Source: Compiled by the author based on own survey data, 2018-19

Higher proportions of the poor and penury farmers had received extension services within a group approach more (in terms of frequency) than the middle-rich and rich farmers. This is because the number and geographical location of the CAs make it easier and economical for the group approach. There were no statistically significant differences in terms of access to extension through media (radio and TVs), on the contrary, higher proportions of middle to rich farmers (50% rich, 77.4% middle-rich and 54.5% middle) had accessed extension services through individual visits between 2 and 6 times in the 2018/19 agricultural season (*Table 7.15*).

These findings have profound implications for my work because the groups approach dominated the lower echelons in order to minimise the costs of extension on the part of the government. This fact is the same rationale that underpins agricultural cooperative input and output marketing, that of the need to reduce transactions costs by aggregating purchasing and marketing activities. This approach of extension delivery to the rural areas (particularly to the CAs) has been in use ever since the attainment of independence and has been the preferred method of extension for the government. Thus, if it is used together with the structures of the cooperative (where more cooperative leaders and members are trained to provide extension services), then win-win situation between the government and the cooperative can be attained.

Access to and utilisation of credit and inputs

Access to agricultural input markets is one of the many problems that affect not only farmers in Zimbabwe's agrarian structure but is pervasive throughout Sub-Saharan Africa (Moyo, 2011a, pp. 953–955; Scoones et al., 2010, p. 101). In cases where markets have been 'functional' (contract farming in tobacco and cotton, outgrower markets in sugar cane production), they have resulted in exploitation of the small-scale farmers much to the negation of the development agenda (Amin, 2012, p. 18; Bernstein, 2005, p. 77; Mazwi & Muchetu, 2015; Scott, 1985, p. 56). Thus, by analysing access and utilisation of inputs, I sought to understand deeply, the challenges and potential solutions that can be forged to overcome the contradictions posed by class differentiation. This process will also help us understand the role that cooperatives can or cannot play within the agricultural input marketing system. As many economists argue, productivity is a direct function of land, labour and capital (Marx, 1992, p. 77). Zimbabwe is one of the countries that have reasonably good quality or strong labour force, and farmers accessed land during the FTLRP, thus securing the 'land' and 'labour' parts of the productivity function. The constraining factor thereof is access to agricultural inputs and capital for equipment. Low access to inputs, brought about by inadequate access to agricultural financing is the reason why the land reform has not achieved its production objectives up to now.

Access to loans and credit

Overall, farmers across the five peasant categories highlighted that their access to credit and financial markets was either average (35.4%), poor (19.3%) or extremely poor (40.6%) (Table 7.16). Only 4.7% of the respondents reported that it was relatively easy to access credit.

Table 7.16: Ease of accessing credit from the cooperative structures.

	Extremely hard		Hard		Average		Easy		Very easy
	No.	%	No.	%	No.	%	No.	%	No.
Penury	7	46.7	6	40	2	13.3	0	0	0
Poor	35	46.1	24	31.6	14	18.4	3	3.9	0
Middle	31	47	4	6.1	31	47	0	0	0
Middle-Rich	4	12.9	2	6.5	19	61.3	6	19.4	0
Rich	1	25	1	25	2	50	0	0	0
Total	78	40.6	37	19.3	68	35.4	9	4.7	0
Pearson Chi-Square Tests: Chi-square = 60.603; df = 12 Sig. = .000*									
Results are based on nonempty rows and columns in each innermost sub-table.									
* The Chi-square statistic is significant at the .05 level.									

Source: Compiled by the author based on own survey data, 2018-19

However, data from the questionnaire reviewed statically significant differences (at 5% CI) between cluster access to credit. For example, more than 50% of the farmers in the penury and poor echelons listed the process to be hard or extremely hard (Table 7.16). Higher proportions (61.3% middle-rich and 50% of the rich cluster) rated the ease of accessing credit as average as we go up the class hierarchy.

The poor and penury farmers, who have no such collateral area are automatically left out of the agricultural finance markets. Consequently, this will reduce the level of development given the fact that Zimbabwean agrarian structure is ridiculed with poor, penury and middle peasants. Effective alternatives sources of credit are therefore needed. Recent literature has suggested that post the FTLRP, rural finance has mostly taken the form of private micro-finance at usury rates (Vitoria et al., 2012).

The approach used in these private-led micro-finance lending initiatives is predominantly group lending in which small loans are given to a group of farmers to engage in farming activities, instead of being given to individuals. Thus, the whole group is held accountable in the case in which one member fails to re-pay their debt. This way, farmers encourage each other to be productive, and they are forced to pass agricultural information that improves the output of each member of that lending group. This is yet another example of the cooperative system in action to provide loans to farmers with little or no collateral. Thus, when the markets underserve the poor farmers, cooperatives are critical in filling the gaps.

Many of the farmers (63% across all clusters) in my survey had never tried to borrow from financial institutions like banks and lending companies. One farmer highlighted that it was even useless to try because they felt that banks would decline their applications. The minority that tried to borrow had not experienced any significant challenges (20.8%), and only about 16.1% of them had significant challenges in accessing credit. Although higher proportions of farmers in the penury to middle clusters had never tried to borrow finance, and also those who tried had experienced many challenges, the difference between clusters was not statistically significant (*Table 7.17*). This implies that the problems in accessing credit affects all the farmers across the clusters and access to finance are, therefore, a critical component of the contemporary AQ. One concern also is the fact that 14 out of the 15 farmers who belong to the penury cluster had never tried to borrow. Such a result supports Chayanov's theory of peasant cooperatives which suggest that the lowest peasant class are oriented towards zero on-farm production and that they solely rely on income from the sale of their labour (or on other sources of finance such as remittances; see page 297). Thus, they usually are not concerned with trying to borrow any credit for agricultural production.

Additionally, due to the reduced flow of information within this cluster, these farmers do not know where and how to apply for production credit. In such a case, under a cooperative system, these farmers can be provided with both information and with credit such that they can produce and move up the ladder. Amongst the farmers that experienced challenges in accessing credit, 83.9% of them cited lack of collateral-related issues to secure the loans from banks. This finding does not come as a surprise given the long-standing disagreement between the government and the banking sector to recognise the state land tenure documents as proof of tenure for collateral (see Moyo et al., 2014).

282

Table 7.17: Presence and nature of challenges in accessing credit for agricultural production.

Peasant cluster	No challenges		Never tried to borrow		Faced challenges		Nature of challenges					
							No collateral security		Unaware of credit facilities		Failure to meet other bank requirements	
	No.	%	No.	%	No.	%	No.	%	No.	%	No.	%
Penury	0	0	14	93.3	1	3.2	1	100	0	0	0	0
Poor	21	27.6	39	51.3	17	54.8	14	82.4	1	5.9	2	11.8
Middle	10	15.2	45	68.2	10	32.3	9	90	1	10	0	0
Middle-Rich	8	25.8	21	67.7	2	6.5	1	50	0	0	1	50
Rich	1	25	2	50	1	3.2	1	100	0	0	0	0
Total	40	20.8	121	63	31	16.1	26	83.9	2	6.5	3	9.7
Pearson Chi-Square Tests: Chi-square = 14.548 df = 8 Sig. = 0.769												
Results are based on nonempty rows and columns in each innermost sub-table.												

Source: Compiled by the author based on own survey data, 2018-19

Some farmers never experienced any challenges in accessing credit, and they eventually got credit or loans from banks and other institutions. For the period from the 2012/13 to 2017/18 agricultural seasons, approximately 33.9% had unpaid debts to various institutions. On average, each of these households had an outstanding loan of US$ 212, with an average interest of 10% per annum. Ironically, the bulk of this money (63.1%) was owed to the cooperative credit facilities followed by that owed to friends and family (20%) and then to local lenders (16.9% to micro-finance schemes) (see *Table 7.18*). The poor (76%) and the middle-rich clusters (69.2%) had the highest proportion of farmers that owed the cooperative credit scheme than any other cluster. Cooperative members were not repaying their debt to the cooperative due to the weak debt collection mechanism, as discussed on page 267. Cooperative structures lack a robust mechanism to deal with issues of corruption and bad debtors. Local lenders and micro-finance institutions have been known to confiscate assets in times of bad debts which the cooperative doesn't do.

On the other hand, borrowers stand to lose a lot more if they do not repay debts to family and friends through strained social relationships than they do if they do not repay the debt to the cooperative. Thus, robust measures need to be put into place to reduce the occurrence of bad debtors within the cooperative movement. Within the clusters, the debt was taken and split between education and on-farm production for the penury peasants while poor, middle and middle to rich clusters channelled the bulk of the funds towards agricultural production.

Table 7.18: Amount, interest, source and purpose of overdue debts in the 2017/18 season

		Penury	Poor	Middle	Middle-Rich	Rich	Total
Amount (US$)	Mean	308	211	230	170	180	212
Interest (%)	Mean	5	8	12	10	0	10
Owed to who (% of farmers)	Friends	50	20	16.7	15.4	100	20
	Local lender	0	4	33.3	15.4	0	16.9
	Coop credit scheme	50	76	50	69.2	0	63.1
Borrowed for what purposes (% of farmers)	On-farm production	50	80	83.3	84.6	0	80
	Education	50	12	12.5	7.7	100	13.8
	Pay off other debts	0	8	4.2	7.7	0	6.2

Source: Compiled by the author based on own survey data, 2018-19

All of the rich peasants who took out loans (from family and friends) directed these loans to payment of school fees for their children (education). Again, this data seems to validate the theory of peasant cooperative, which suggest that the rich peasants are not concerned with trying to maximise profits from agricultural activities. This is because their primary source of income is outside agricultural production (further discussion from page 297).

Access and utilisation of inputs

In addition to poor access to credit and finance in the agrarian structures of Africa, various scholars argue that development in Africa lags behind other regions due to inadequate access to Green-Revolution-type of inputs (Eicher, 1995, pp. 805–807). These include fertilisers, hybrid seeds and chemicals. However, political economists criticise the overreliance on Green-Revolution type inputs because it displaces indigenous peasant seed production systems with multinational companies' "suicide" hybrid systems as witnessed in Asian countries in the mid-1900s (De Janvry, 1981, pp. 124–125; Patnaik et al., 2011, pp. 75–76; Scott, 1985). Suicide hybrid seeds[19] render production highly dependent on these inputs providers because the seeds cannot be retained for more than one season. Overall, I found that use of certified, patented seed and inorganic fertiliser was as high as 73.4% and 70.3% of all the households during the 2017/18 agricultural season respectively. However, an average of around 67% of the penury, poor and middle clusters used these inputs (*Figure 7.9*). Use of herbicides and pesticides was generally higher for upper-end clusters than it was for lower end. Only a few of the surveyed households utilised retained seed and livestock mineral licks, while everyone across the clusters accessed and utilised livestock feed and drugs.

A steady increase in the proportion of farmers that used inorganic fertilisers from penury to rich peasant clusters, and the opposite was correct regarding the deployment of organic fertilisers. Organic farming, as we saw in Chapter Five for Japanese organic cooperatives, is an alternative method of farming that reduces the use of harmful inputs, like pesticides, by encouraging natural methods of weed and pest control. Organic farming can be done efficiently within the confines of a cooperative in which each member helps and monitors one another in the use and practice of organic farming.

Save for such things as manure and retained seed (generated within the production cycle), the bulk of the inputs were obtained from agro-dealers stationed within the area. Specifically, 62.2%, 78.8%, 58.5%, 91,1%, 65.6% and 65.1% of the farmers accessed fertiliser, pesticides, certified seed, post-harvest loss (PHL) insecticides, livestock feed and livestock drugs from local agro-dealers respectively (see Figure 7.10). Rural area agro-dealers charged up to three times the wholesale price of agricultural commodities. This means that a farmer who accessed grain protectants from the

---

[19] This is also known as Genetic Use Restriction Technology (GURT) and is the method of restricting use of hybrid patented seeds through activation of some genes making second generation seeds (retained seed) infertile (Jefferson et al., 1999, p. 13)

agro-dealers faced increased overall costs of production (Govereh et al., 2019) than they would if they were buying their inputs in bulk from the whole or the manufacturer.

Figure 7.9: Access and utilisation of agricultural inputs (2017/18')

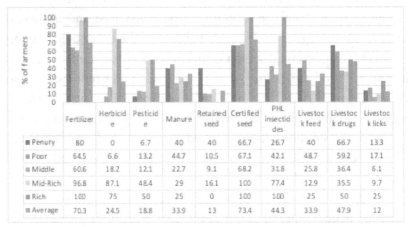

	Fertilizer	Herbicide	Pesticide	Manure	Retained seed	Certified seed	PHL insectid des	Livestoc k feed	Livestoc k drugs	Livestoc k licks
Penuty	80	0	6.7	40	40	66.7	26.7	40	66.7	13.3
Poor	64.5	6.6	13.2	44.7	10.5	67.1	42.1	48.7	59.2	17.1
Middle	60.6	18.2	12.1	22.7	9.1	68.2	31.8	25.8	36.4	6.1
Mid-Rich	96.8	87.1	48.4	29	16.1	100	77.4	12.9	35.5	9.7
Rich	100	75	50	25	0	100	100	25	50	25
Average	70.3	24.5	18.8	33.9	13	73.4	44.3	33.9	47.9	12

Source: Compiled by the author based on own survey data, 2018-19

Some farmers were contracted to produce such crops as tobacco and had to access inputs from the contracting companies. Of particular importance to this book is the role of cooperatives in input purchasing and supply. Approximately 19.7%, 15.7% and 64.7% of the farmers obtained their livestock feed, livestock drugs and mineral licks from the cooperative structures, respectively. The use of this channel depended on the type, and product orientation of the cooperatives, e.g., farmers from non-crop producing cooperatives such as Chikwaka dairy, Gosha egg and Gutu golden egg accessed their inputs from the cooperative. This finding has profound implications for the nature and character of the cooperatives that can work in Zimbabwe.

Figure 7.10: Sources of agricultural inputs (2017/18)

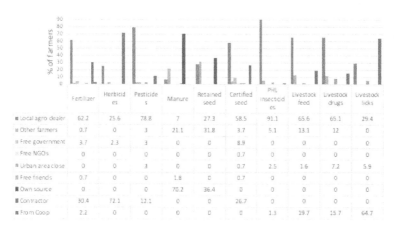

	Fertilizer	Herbicides	Pesticides	Manure	Retained seed	Certified seed	PHL insecticides	Livestock feed	Livestock drugs	Livestock licks
▪ Local agro dealer	62.2	25.6	78.8	7	27.3	58.5	91.1	65.6	65.1	29.4
▪ Other farmers	0.7	0	3	21.1	31.8	3.7	5.1	13.1	12	0
▪ Free government	3.7	2.3	3	0	0	8.9	0	0	0	0
▪ Free NGOs	0	0	0	0	0	0.7	0	0	0	0
▪ Urban area close	0	0	3	0	0	0.7	2.5	1.6	7.2	5.9
▪ Free friends	0.7	0	0	1.8	0	0.7	0	0	0	0
▪ Own source	0	0	0	70.2	36.4	0	0	0	0	0
▪ Contractor	30.4	72.1	12.1	0	0	26.7	0	0	0	0
▪ From Coop	2.2	0	0	0	0	0	1.3	19.7	15.7	64.7

Source: Compiled by the author based on own survey data, 2018-19

An average of 80% of the farmers managed to use either their proceeds from agricultural production or their savings from non-farm income to finance production. Indeed, the bulk of the income used to pay for inputs comes from agricultural proceeds (*Figure 7.11*). These findings provide scope for the success of the cooperative credit schemes that I proposed in Chapter Eight. Contract farming came third in terms of importance as a source of agricultural inputs. Contract farming has been proved to be a viable source of rural financing, one that was able to sustain cotton production between the years 2000 and 2015. However, contract farming has tended to focus on cash crop production much to the neglect of other crops. If contract farming can be directed at food production as attempted by Command Agriculture, then this method of financing can work even within a cooperative framework where the cooperative becomes the contracting company. It is possible to contract farmers to grow maize instead of cotton because maize is more rewarding per hectare than cotton or tobacco, as argued in Muchetu (2019a, p. 51).

In addition to the amount of capital and the levels of inputs accessed, I analysed the levels of labour use in my sample because labour use patterns are critical in defining and describing the nature and character of producers (Patnaik, 1988, p. 302).

Figure 7.11: Sources of income to purchase agricultural inputs (2017/18)

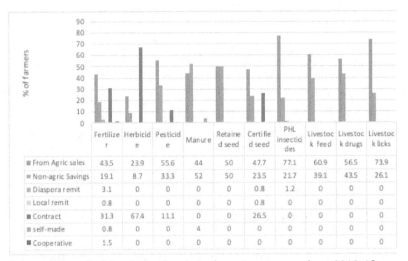

	Fertilizer	Herbicide	Pesticide	Manure	Retained seed	Certified seed	PHL insecticides	Livestock feed	Livestock drugs	Livestock licks
■ From Agric sales	43.5	23.9	55.6	44	50	47.7	77.1	60.9	56.5	73.9
■ Non-agric Savings	19.1	8.7	33.3	52	50	23.5	21.7	39.1	43.5	26.1
■ Diaspora remit	3.1	0	0	0	0	0.8	1.2	0	0	0
■ Local remit	0.8	0	0	0	0	0.8	0	0	0	0
■ Contract	31.3	67.4	11.1	0	0	26.5	0	0	0	0
■ self-made	0.8	0	0	4	0	0	0	0	0	0
■ Cooperative	1.5	0	0	0	0	0	0	0	0	0

Source: Compiled by the author based on own survey data, 2018-19

As I explained earlier, labour use was one of the variables that helped us classify these agricultural producers. I found no statistically significant differences in self-employment traits between the clusters as most of these used own-labour on their farms (an average of 3.6 persons per farm). Here, the data I found deviated from the confines of Chayanov's theory of cooperatives which argued that the uppermost echelons within the peasant classes hardly used own-labour on their farms. This is due to the fact that our study only collected the number of persons employed on the farm and did not go more in-depth into the quantity of labour invested by the persons in terms of labour days as done in studies such as Chambati (2017). Thus, a 'reduced' indicator of labour was used, and no significant radical picture could be found.

Rich peasants had the highest average (4.3 persons) working on their farm (*Figure 7.12*). This can be explained, chiefly by the fact that labour cost accounts for a significant amount of production cost, and rich peasants would not want to increase the amount of cost of agricultural production (which is not really their primary source of income) and would instead utilise own-labour in conjunction with power-driven machinery. This seems plausible and also explains the situation obtaining in the Japanese rich peasants who rely entirely on family labour and machinery (see Chapter Five).

The same situation was obtaining in the utilisation of permanent hired labour which averaged 2 in the rich cluster, 0.5 in the penury, 0.7 in the middle-rich and almost none in the rest of the clusters. In this case, the rich lived up to the expectations of the Chayanovian theory; however, the penury or the lowest cluster did not; instead, they performed better than the poor peasant cluster which hired an average of 0.1 persons per household. This points to a contradiction that although farmers would sell off their labour, one in every two of such households had a permanent worker. This is so because some farm owners can decide to provide labour to another well up farmers in order to get inputs or draught power. This is particularly true for those A1 and CA farmers that are surrounded by A2 commercial farmers who usually own power-driven tractors and other machinery. Although the farmer in the penury cluster is selling their labour (2 persons on average) and hiring others, differences in wages exist between A1, CA and A2 commercial farmers (see Chambati, 2017).

Figure 7.12: Average labour access and utilisation (2014-2018)

	Penury	Poor	Middle	Middle-Rich	Rich	Total
Family labour hired-in	3.8	3.5	3.5	3.8	4.3	3.6
Permanent hired labour	0.5	0.1	0.4	0.7	2.2	0.4
Casual hired labour	0.8	1.1	1.3	10.3	16.8	3
Family labour sold-out	2	0	0	0	0	0.2

Source: Compiled by the author based on own survey data, 2018-19

There were statistically significant differences in the casual labour hiring behaviours of the five different clusters. The number of seasonal workers significantly increased from lower penury cluster

289

(0.8 persons) to the rich peasant clusters (16.8 persons). Supplementary investigations into the reasons why this may be so showed that this was due to the changing forms of wage labour witnessed on a global scale (labour is increasingly being casualised within the context of flexible labour markets).

Capitalist entrepreneurs are moving away from hiring permanent labour, which is costlier, to hiring seasonal labour which they can engage with only in times of need and let them go when there is no work (also the seasonality of some agricultural crops that imply that farmers cannot employ labour throughout the year). These findings also reveal the presence of quality labour that straddles between different farming sectors in order to find work (Chambati, 2017).

Agricultural production, output markets and the role of the cooperative

I have established that farmers, especially in the resettled areas accessed land, have relatively adequate labour, but had low access to inputs. This section tries to understand production levels and access to output markets as well as the current and potential role that the cooperative can play in the process. Although the cooperatives were focused on agricultural commodities such as eggs, milk and market gardening, and irrigation equipment sharing, an overwhelming number of the individual farmers were involved in non-cooperative maize production across all peasant clusters (92.7%). Thus, farmers try to attain food security through own-production and only venture into other income-generating activities to purchase goods and services that they cannot produce on their own. This fact has severe implications for the construction of a cooperative model because such an enterprise need not only focus on the production of a specific commodity (e.g., eggs) without factoring in the time that farmers will dedicate to maize production otherwise it will most likely fail.

There were statistically significant differences in the cropped area under maize as I moved from penury to rich clusters. The penury and the poor clusters put on average, half a hectare of their field under the crop, while the middle-cluster averaged one hectare, and the middle-rich and the rich averaged 2 hectares between 2015 and 2018. The second most produced crop outside the focus of their

290

cooperatives was tobacco, grown on an average of 1.18 ha per farmer (higher in the rich-cluster). However, there were differences in the proportions of farmers that were producing the cash crop as evidenced by only 5 in 15 of the penury farmers and 39 of the 76 poor farmers (Table 7.19). Approximately 61.3% of the and 75% of the middle-rich and rich grew tobacco, respectively.

Table 7.19: Area under primary crop production

Peasant cluster	N	Maize			Soybean			Tobacco			Groundnuts		
		Pp^1	$\%Pp^2$	Mn^3	Pp^1	$\%Pp^2$	Mn^3	Pp^1	$\%Pp^2$	Mn^3	Pp^1	$\%Pp^2$	Mn^3
Penury	15	14	93.3	0.56	0	0.0	-	5	33.3	0.86	5	33.3	0.44
Poor	76	70	92.1	0.50	8	11.4	0.99	29	38.2	0.81	31	40.8	0.28
Middle	66	62	93.9	1.03	6	9.7	0.98	27	40.9	1.05	25	37.9	0.55
Middle-Rich	31	28	90.3	2.62	5	17.9	2.27	19	61.3	1.84	14	45.2	0.46
Rich	4	4	100.0	1.97	1	25.0	3.05	3	75.0	2.12	2	50.0	0.58
Total	192	178	92.7	1.06	20	11.2	1.41	83	43.2	1.18	77	40.1	0.42

N=Total number of farmers in the sample; Pp^1=Number of crop producers; $\%Pp^2$= Proportion of crop producers to the total number of farmers; Mn^3= Mean area under crop
Source: Compiled by the author based on own survey data, 2018-19

An appreciation of the dynamics of cropped areas and the proportion of farmers engaged in the production of these crops will be vital in the proceeding sections of this book. This is because it helps us understand which crops are essential to the farmers, and hence, the type of farming services that a cooperative would have to provide. The least grown major crops were groundnuts and soybean, which averaged 40.1% and 11.2% of the farmers and were put under 0.42ha and 1.41ha of land, respectively. Groundnuts are usually grown for subsistence purposes while soybean is for commercial purposes.

These cropping patterns, however, need to be understood within the concept of land utilisation. Media outlets and several other opponents of the FTLRP often cite land under-utilisation as the reason why the reform had failed (Bond, 2008; Gumede, 2018;

Johnson, 2009). However, my survey seems to support various other scholars who have reported increased levels of the cropped area from the pre-FTLRP era. On average, farmers had 43.7% of their land under some crops. Most of the land was dedicated to maize production (76.8% of the land, highest in the penury clusters), followed by soybean (40.4), then tobacco (40.1%) and groundnuts (17.7%) (*Figure 7.13*). Cooperative activities were being done on tiny portions of their land as a sideshow activity for most of the cooperatives. Since the bulk of the land had non-cooperative crops like maize and tobacco.

Figure 7.13: Land utilisation rates by major crops grown (2017/18)

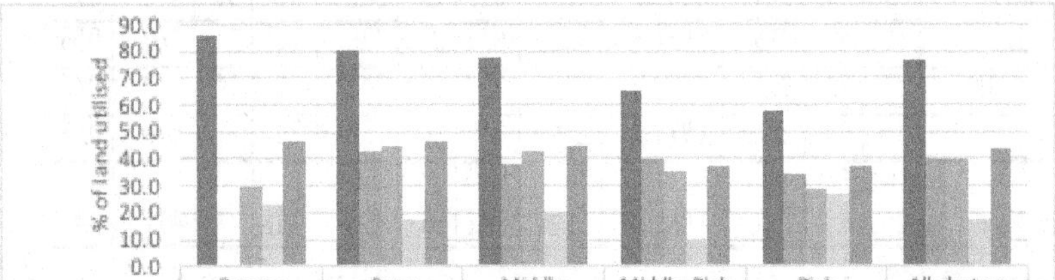

% of land utilised	Penury	Poor	Middle	Middle-Rich	Rich	All clusters
Maize	86.0	80.3	77.5	65.1	57.3	76.8
Soybean	0.0	43.0	38.1	39.5	34.6	40.4
Tobacco	29.7	44.7	42.3	34.9	29.0	40.1
Groundnuts	22.7	17.3	20.6	10.4	26.7	17.7
Average	46.1	46.3	44.6	37.5	36.9	43.7

Source: Compiled by the author based on own survey data, 2018-19

Land utilisation, across all crops, was highest among the lower clusters because most of the farmers in the lower echelons hail from the CAs where the average land-holding is less than one hectare. Thus, farmers will cultivate all their land with some even renting in.

There are some cases where CA farmers rent in the farming area from the nearby A1 farms, and some A1 farmers that rent in from A2 farms. The bulk of the farmers in the middle-rich and rich-echelons were A1 farmers (accessed four to six hectares of arable land). In the current economic environment, where agricultural finance and inputs are scarce, putting all the land under agricultural production is difficult, hence the lower levels of land utilisation recorded amongst the surveyed households. To ascertain whether land utilisation translates to higher income and productivity levels, I examined the output levels in monetary terms.

It was not surprising to note that on average, higher-end clusters performed better than lower-end ones. On average, from agricultural production alone, farmers in the penury class obtained a gross value of output worth US$825 per annum from their agricultural production activities, compared to US$1,175 and US$910 for the poor and middle clusters respectively (*Table* 7.20). This result was in stark contrast with the middle-rich and the rich peasants who obtained US$4,366 and US$12,898 respectively. Thus, with these amounts of earnings from agricultural activities alone, I maintain the position that rural households can be productive enough to sustain themselves and self-finance their next agricultural activities. This is particularly so given that the food poverty line (FPL) and the Total Consumption Poverty Line (TCPL) for 2017 stood at US$2,200 and US$6750 for the average Zimbabwe family of five respectively (ZimStats, 2018). Higher class peasants were making money through the cooperative within the Zimbabwe agrarian Cooperatives.

As it is, money earned in US dollars from agriculture is usually stored in a bank (farmers are paid through bank transfers for tobacco and soybean production, some of the money for maize is also paid through mobile money channels such as Ecocash). Farmers can only use it (local currency) when they want to purchase something. What this means is that banks make more money and farmers are not paid anything for their contributions, and when they ask for loans, they are told that they need to produce collateral. Cooperatives schemes, which can grow into cooperative banks, can capture and solve this problem by giving some of the money back to the farmers.

Across all the clusters, higher income was made from agricultural activities outside the cooperative because some of these cooperatives were just for irrigation and did not care about any other activities that a farmer was involved in, thus calculating the amount of production dedicated to the cooperatives was a considerable hurdle. Additionally, a number of the farmers joined the cooperatives, such as egg production and market gardening as a second source of income, or as an off-season activity since their primary crop production is rainfed.

Table 7.20: Production levels: Income from agricultural production (USD)

	Agric prod under the coop			Agric prod not under the coop			Overall	
	No.	%	Mean	No.	%	Mean	No.	Mean
Penury	14	93.3	292	4	26.7	1088	15	825.47
Poor	60	78.9	522	31	40.8	1522	76	1175.07
Middle	41	62.1	1039	24	36.4	728	66	909.98
M-Rich	28	90.3	3849	16	51.6	2217	31	4366.16
Rich	3	75.0	11863	2	50.0	8000	4	12897.5
Total	146	76.0	1516	77	40.1	1564	192	1816.08

Source: Compiled by the author based on own survey data, 2018-19

However, on average, 76% of the farmers were engaged in cooperative related activities, while 40.1% had other extra-cooperative production activities (see *Table* 7.20). This was a crucial finding in light of the results that I learned from the Chiba cooperative in Japan, where some farmers had stopped selling their products through the cooperative structures altogether.

There were three dominant marketing channels that farmers used to sell their crops to the market. For the penury (60%), poor (73.7%) and middle (34.8%) households, who were mostly involved in cooperative oriented production, the cooperative channel was one of the viable marketing channels they utilised (Figure 7.14). Middle-rich and rich households utilised other channels beyond the cooperatives. A considerable number of farmers also sold their output on the local markets, which are known to have lower output prices within the agricultural marketing system (Muir-Leresche & Muchopa, 2006; SMAIAS, 2015).

These results point to the fact that agricultural cooperatives in Goromonzi, or Zimbabwe, are now focusing more on selling output and have neglected other functions such as input supply which they were previously known for (CACU chairperson interview, 2018). This a useful piece of information for the construction of the model in the proceeding chapters.

Figure 7.14: Top three most-utilised marketing channels by cluster (2017/18)

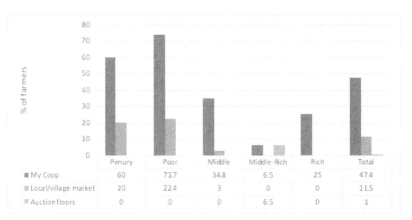

	Penury	Poor	Middle	Middle-Rich	Rich	Total
■ My Coop	60	73.7	34.8	6.5	25	47.4
■ Local/village market	20	22.4	3	0	0	11.5
■ Auction floors	0	0	0	6.5	0	1

Source: Compiled by the author based on own survey data, 2018-19

It is imperative to note at this stage that the type of cooperative and the type of commodity influenced the choice of marketing channel. Dairy and egg cooperative, which requires sophisticated machinery during the transportation and movement of the milk & eggs, had more farmers selling their output through cooperative structures (see *Figure 7.15*). The cooperatives were established for this reason.

Other cooperatives, such as Kumboedza, Shungu, Simba Ivhu and Survival Skills, used a combination of cooperative and local market channels. This is also because of the nature of their products, which is not fragile and does not require specialised transport vessels as in the case of milk and eggs. On the other hand, Xanadu A (a tractor and equipment sharing cooperative) and Tagarika (irrigation equipment sharing cooperative) had meagre proportions of farmers who had utilised any of the channels. These farmer's commodity marketing behaviour proved challenging to capture because they only met or cooperated as far as sharing equipment was concerned.

Figure 7.15: Top three most utilised marketing channels by cooperative (2017/18)

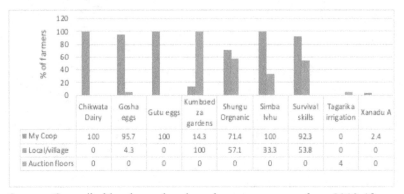

	Chikwata Dairy	Gosha eggs	Gutu eggs	Kumboed za gardens	Shungu Organic	Simba Ivhu	Survival skills	Tagarika irrigation	Xanadu A
■ My Coop	100	95.7	100	14.3	71.4	100	92.3	0	2.4
■ Local/village	0	4.3	0	100	57.1	33.3	53.8	0	0
■ Auction floors	0	0	0	0	0	0	0	4	0

Source: Compiled by the author based on own survey data, 2018-19

Higher prices for the output were mentioned by the majority of farmers as the key reason for choosing a marketing channel to sell their products. This was true for all the clusters. However, an interesting finding was the fact that some farmers in the penury cluster (33.3%) reported that it was a statutory requirement for them to sell their products through the cooperative. On average, 13.5% of the farmers reported that they had to sell through the cooperative structures. The question of whether the cooperative allow their members to sell their produce to other cooperatives or just be restricted to the cooperative has illuminated cooperative management debates in Japan (Godo, 2015, 2016).

In Japan, cooperatives had to abide by the Anti-Monopoly Act, which allows farmers or members of a cooperative to exercise their discretion when choosing a marketing channel for their produce.

I analysed this further and found that two cooperatives, in particular, had higher proportions of farmers who believed that they had to sell their output through the cooperative. These were Chikwaka dairy cooperative (65.5%) and Survival Skills cooperative (38.5%) as shown in *Figure 7.16*. The former is a milk production cooperative, and thus it could be understood that milk had to be sold through the cooperative in order to maintain such things as food safety, quality, and standard. While the latter is a market gardening cooperative that producers horticultural produce which has more flexible marketing options. Survival Skills wanted to maintain standards and meet minimum output quantities that would lower their transactions cost. Farmers in this cooperative also reported that the price for the product was a pulling factor as well. All other cooperatives members from other cooperatives highlighted that they considered the price of the output under offer before deciding on a channel.

Figure 7.16: Reasons for selecting marketing channel by cooperative (2017/18)

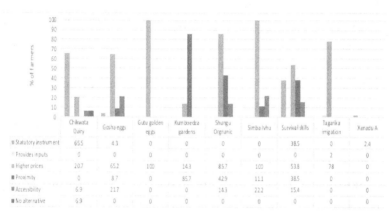

	Chikwata Dairy	Gosha eggs	Gutu golden eggs	Kumboedza gardens	Shungu Orgnanic	Simba Ivhu	Survival skills	Tagarika irrigation	Xanadu A
Statutory instrument	65.5	4.3	0	0	0	0	38.5	0	2.4
Provides inputs	0	0	0	0	0	0	0	2	0
Higher prices	20.7	65.2	100	14.3	85.7	100	53.8	78	0
Proximity	0	8.7	0	85.7	42.9	11.1	38.5	0	0
Accessibility	6.9	21.7	0	0	14.3	22.2	15.4	0	0
No alternative	6.9	0	0	0	0	0	0	0	0

Source: Compiled by the author based on own survey data, 2018-19

Incomes and accumulation trajectories

The study has discussed the production behaviours of the cooperative members as well as the various decision-making processes that take place during output marketing. The examination of whether gross incomes derived from agricultural production were enough to sustain their lives and access to other sources of income extends henceforth. These questions are vital in the understanding of the peasantry, considering the base theoretical framework because they aid us in identifying and differentiating farmers when modelling their role in the cooperative framework. In Chayanov's analysis of the peasantry in Russia, a very few peasants had access to other non-agricultural sources of income. Although there are many differences in both time and space for Russia and Zimbabwe peasants, Chayanov's theory of peasant cooperatives can still be utilised as a base analytical framework. The next step is to get an in-depth understanding of income diversification in Goromonzi, and how these shape their respective accumulation trajectories.

Incomes

The two most reported sources of income for the respondents were agricultural production activities followed by remittances received. Overall, 79.7% of the households across the five clusters had received some money income from agricultural activities done under the cooperative while 40.6% had other agricultural activities that had nothing to do with the cooperative (*Figure 7.17*). Although there were no significant statistical differences between clusters who received incomes from cooperative related activities, in the case of non-cooperative related income sources, the upper or richer-clusters of the peasantry had higher proportions of farmers earning income outside the structures of the cooperative.

This finding has two implications: the first one is that the middle to rich clusters is able to stand on their own, and hence it may be difficult for them to stay in the cooperative. However, the second implication is that these farmers are able to stand on their own using alternative incomes but still choose to be confined to the structures of the cooperative, this result reflects on the functionality and attractiveness of the cooperative (rich clusters were also making profits from the cooperative business). It means that even the rich can see or realise benefits using the cooperative channels of marketing.

Approximately 29.7% and 16.1% of the farmers had received local remittances and diaspora remittances, respectively. The poor-cluster (39.5% of the farmers) dominated local remittances, while the middle-rich (25.8%) dominated diaspora remittances.

Given the inaccessibility to finance capital, these alternative sources of finance have been instrumental in bridging the rural agricultural funding gap. Other sources of finance were petty trading (11.5%) and sell of labour (8.3%). Of interest in this finding are the overwhelming proportions of rich-cluster peasants involved in petty-trading and that of penury peasants in the sale of labour (*Figure 7.17*). Again, this seems to validate Chayanov's theorisation of the

298

peasantry since the rich mainly derive their incomes from investments outside agriculture while the lowest echelon derives its income from the sale of its labour to other farmers (Chayanov, 1991).

Figure 7.17: Sources of income for cooperative members (2016-2018)

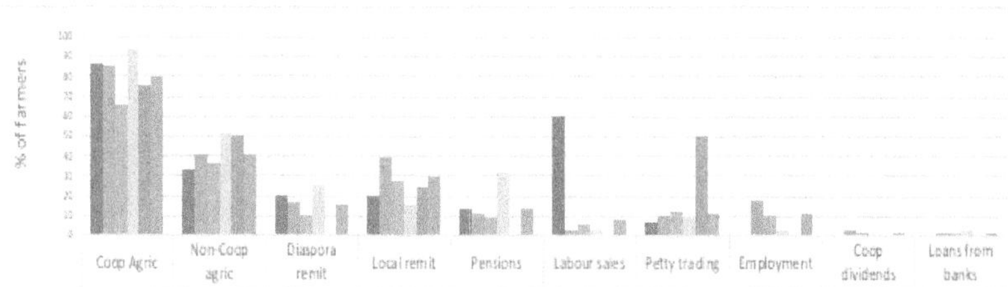

	Coop Agric	Non-Coop agric	Diaspora remit	Local remit	Pensions	Labour sales	Petty trading	Employment	Coop dividends	Loans from banks
Penury	86.7	33.3	20	20	13.3	60	6.7	0	0	0
Poor	85.5	40.8	17.1	39.5	11.8	2.6	10.5	18.4	2.6	1.3
Middle	65.2	36.4	10.6	27.3	9.1	6.1	12.1	10.6	1.5	1.5
Middle-Rich	93.5	51.6	25.8	16.1	32.3	3.2	9.7	3.2	0	3.2
Rich	75	50	0	25	0	0	50	0	0	0
Total	79.7	40.6	16.1	29.7	14.1	8.3	11.5	11.5	1.6	1.6

Source: Compiled by the author based on own survey data, 2018-19

There were significant statistical differences between the average amount of income obtained from various sources within the clusters. The upper-echelons (rich and middle-rich) dominated in terms of incomes derived from the two forms of agricultural production (local remittances and petty-trading).

The middle-cluster enjoyed higher average incomes from wages and salaries in formal employment the middle-rich (and penury clusters) received higher average receivables from pensions, while other farmers in the penury and poor clusters were also engaged in brick moulding (US$525 per annum) (*Figure 7.18*).

The fascinating point from these findings is the fact that agricultural production continues to be a significant source of income in terms of the total number of people engaged in the activity as well as the absolute values/amounts of earnings from these livelihood strategies. This result gives scope for the formation and possible sustainability of a viable agricultural cooperative system centred around one agricultural product. If farmers depended on off-farm income sources, for instance, the formation and development of such cooperative organisations would be implausible.

Figure 7.18: Average income earnings from various sources (2017)

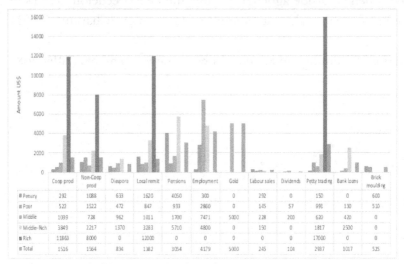

	Coop prod	Non-Coop prod	Diaspora	Local remit	Pensions	Employment	Gold	Labour sales	Dividends	Petty trading	Bank loans	Brick moulding
▪ Penury	292	1088	633	1620	4050	300	0	292	0	150	0	600
▪ Poor	522	1522	472	847	933	2860	0	145	57	991	130	510
▪ Middle	1039	728	962	1011	1700	7471	5000	228	200	620	420	0
▪ Middle-Rich	3849	2217	1370	3283	5710	4800	0	150	0	1817	2500	0
▪ Rich	11863	8000	0	12000	0	0	0	0	0	17000	0	0
▪ Total	1516	1564	834	1382	3054	4179	5000	245	104	2937	1017	525

Source: Compiled by the author based on own survey data, 2018-19

Accumulation

Twenty years after the FTLRP, it certainly makes sense to try and analyse the accumulation patterns of farmers in the resettled areas because they are past their 'farm establishment' stage of farm development. In this sub-section, I analysed the ownership of production assets and equipment as well as that of productive and non-productive on-farm. I tried to understand how this could affect the ability of the respective farmers to form and join the cooperatives.

Higher proportions of farmers possessed hand tools, followed by those that owned animal-driven implements while a few farmers owned power-driven implements such as vehicles and tractors. As is expected, higher-end clusters dominate power-driven assets ownership, and the penury and poor dominate the ownership of hand-held implements. However, what is most interesting is a comparison of my results with those obtained in the Sam Moyo African Institute for Agrarian Studies (SMAIAS) household survey carried out in 2014/15. The survey covered six districts across the country, including the Goromonzi district.

Table 7.21: Percentage ownership of productive assets in Goromonzi, 2011 & 2018

	SMAIAS survey 2011 (N=1090)			Own Survey 2018 (N=192)		
	A1	CA	Total	A1	CA	Total
Hoes	93.4	98	95.5	100	100	100
Axes	90.2	92	91	95.7	88	91.7
Wheelbarrow	67.2	62	64.9	71.7	59	65.1
Spade folks	41	12	27.9	81.5	60	70.3
Knapsack	77	16	49.5	77.2	53	64.6
Animal drawn ploughs	60.7	54	57.7	35.9	51	43.8
Scotch carts	44.3	38	41.4	19.6	37	28.6
Generator	16.4	4	10.8	14.1	11	12.5
Tractor	3.3	0	1.8	8.7	0	4.2
Vehicles	1.0	0.5	1.6	21.7	5	13
Water pump	1.5	0.6	1.9	30.4	2	15.6
Grinding mills	0.6	0	0.8	2.2	0	1

Source: Compiled by the author based on own survey data (2018/19) and SMAIAS household survey data (2014/15)

Slightly higher proportions of farmers in my survey had reported owning productive assets than did those in the SMAIAS survey in 2011 (see *Table 7.21*). Further analysis showed that higher proportions of farmers had accumulated hand tools and power-driven tools, while the proportion of farmers who had bought animal-driven tools had reduced. Although more data is needed to make more robust inferences, this is a positive development towards a preference for more technologically advanced methods of farming in which power-driven ones slowly replace animal-drawn implements. More in-depth analysis is necessary to ascertain the effect on asset accumulation of their participation in the cooperative; the current data could not establish this because farmers struggled to separate sources of income and the respective expenditure.

The same situation was obtaining in the investment by farmers in infrastructural assets such as housing, storage, and crop processing facilities. Indeed, accumulation was happening in the households despite the adverse economic environment and depressed levels of financial market accessibility. On average, higher proportions of people had invested in deep-wells in my survey (78.6%) as compared to the proportion of farmers in the SMAIAS survey (54%) (see *Table 7.22*).

In their survey report, SMAIAS (2015, pp. 155–160) argued that there had seen increased asset accumulation in Goromonzi (and all other districts) basing this on earlier baseline survey they had carried out in 2006/7. My survey results show increases in the proportion of farmers who invested their incomes in Blair-toilets and storage facilities as well as in deep-wells (water points). Thus, my study further supports this argument that the accumulation of both productive and non-productive assets is underway in Goromonzi district. This ability of the farmers to accumulate and alleviate themselves from poverty is motivating because it shows that under the right conditions (that can be set by the cooperative or any other rural social organisations), the peasantry can self-finance their way out of under-development.

Table 7.22: Percentage ownership of on-farm infrastructure by peasant cluster (2017/18)

	SMAIAS survey 2011			Own Survey 2018		
	A1	CA	Total	A1	CA	Total
Deep well	48	60	54	63.0	93	78.6
Brick/asbestos/Zinc	21.7	40	30.9	23.9	58	41.7
Private tobacco barns	1.5	5	3.3	6.5	5	5.7
Communal tobacco barns	30	0	0.2	25.0	0	12.0
Irrigation	12	6	9	15	6	12
Blair toilets	15	40	27.5	67.4	92	80.2
Storage	6.1	1.9	4	27.2	0	13.0

Source: Compiled by the author based on own survey data, 2018-19

The farmers utilised among various sources, two primary sources of income to invest in assets and infrastructure: proceeds from agricultural sells (cooperative and non-cooperative production) and savings outside of agricultural production (employment, petty trading, sale of labour).

In general, farmers who bought bigger assets or machinery such as vehicles, tractor and grinding mills had to use incomes derived from personal savings outside agricultural activities. In contrast, with regard to smaller assets such as ox-drawn implements (plough, ridger, planter and harrow), farmers purchased these using income proceeds from agricultural activities (*Table 7.23*).

Table 7.23: Sources of income to fund asset and infrastructure accumulation, 2016-2018

Asset	Proceeds from Agric-sales	Savings outside agriculture	Asset	Proceeds from Agric-sales	Savings outside agriculture
Hand and ox-drawn tools			Power-driven equipment		
Hoes	57.8	42.2	Vehicles	25	75
Axes	62.2	37.8	Generator	50	50
Spades & forks	68.2	31.8	Tractor	0	100
Wheelbarrow	58.3	41.7	Water pump	50	50
Watering cans	59.1	40.9	Grinding mill	0	100
Knap sack	65.4	34.6	Infrastructure		
Scotch-cart	100	0	Deep well	55.6	44.4
Ox-Plough	100	0	Cattle facilities	40	60
Ox-Planter	100	0	Poultry runs	46.2	53.8
Ox-Ridger	100	0	Private barns	50	50
Ox-Cultivator	75	25	Grading shade	66.7	33.3
Ox-Harrow	66.7	33.3	Blair toilets	80	20
Maize sheller	0	100	Storage	66.7	33.3

Source: Compiled by the author based on own survey data, 2018-19

This finding seems to shoot down the argument that farmers cannot accumulate from income derived through agricultural activities. However, we need to understand that savings outside of agriculture are playing the role of lender of the last resort for the bulk of the farmers in an economy that does not provide credit and loans to the farmers. If financial markets were 'working' relatively well, farmers would be able to access loans to purchase machinery and equipment at concessionary rates. They would become more productive (since they will be able to buy inputs, machinery and labour on time) and be able to repay their loans. Thus, the use of savings outside agriculture to purchase farming assets and infrastructure only goes to show the willingness of the farmers to engage in productive farming activities, which is a pre-requisite for the formation of the cooperative system. Again, regarding the trajectory that the national financial markets are taking, more well-up farmers will be able to access credit since they have assets that they can mortgage.

Cooperatives and gender mainstreaming

Gender issues have become topical over the past decade, and rightly so because for so long, women's contribution to agriculture has often been undermined or neglected, and yet they play a significant role in agrarian processes (MAMID, 2012, p. 30; World Bank, 2019a, p. 2). The literature stresses the advantages that cooperatives/local farmer groups give to women, how they improve women's access to productive resources and output markets, hence a key conduit for empowering them. Additionally, cooperatives have been thought of as a last resort for women as seen through the formation of several productive women-only cooperatives throughout Africa and beyond. Women pressure groups to allocate at least 20% of land to women in Zimbabwe's 2000 land reform is an example (Chiweshe, 2011, pp. 41–45; Sachikonye, 1995, pp. 407–408). Thus, my study also assessed the gendered aspects of the cooperative movement to try and understand how women were involved in the movement.

In Zimbabwe, women empowerment has always been depressed, even in the pre-independence era. For instance, women held less than 4% of the land around in the pre-1980 white commercial sector, and their access to other resources was also highly constrained. The situation continued after independence until the FTLRP, which improved women's access to land to a figure between 19.5 and 24% (SMAIAS, 2015; Utete, 2003). Such an improvement has nonetheless come after over more than four decades, and a lot more needs to be done to move towards equitable distribution of land resources since their role in various agricultural production activities exceeds 50% (World Bank, 2019a, p. 2). From my study, higher proportions of female cooperative members were found in the CA movement as compared to that of the resettled areas. This finding can be seen through the aggregate of cooperatives from the CAs (74.9% females), as shown in *Table 7.24*. Most of the cooperatives in the CAs have almost equal representation except in two of the cooperatives, Gutu eggs (with an 82.4% female membership) and Kumboedza gardens which is effectively a woman only cooperative. These results point to the power of cooperatives to attract disadvantaged groups. There virtually is no other productive rural activity that can attract more women than the cooperative.

In the resettled areas, female representation stood at 36.6% of the total membership. Although this figure is lower than that of the cooperatives in the CAs, it is essential to note that land ownership is a pre-requisite for the cooperatives that exist in the resettled areas. Thus, given the fact that female land ownership stood well below 25%, a 36.6% participation rate of females still points to the strength of the cooperative to include women in production activities. Overall, there were more women (65.8%) in the cooperative movement than there were men, which goes on to highlight that cooperatives can answer gender parts of the contemporary AQ.

Table 7.24: Gender composition, by cooperative and by peasant classes

By cooperative	Males	Females	Total	% of females	By peasant cluster	Male No.	%	Female No.	%
Chikwaka	34	31	65	47.7					
Gosha	15	23	38	60.5	Penury	12	9	3	5.2
Gutu	60	280	340	82.4	Poor	55	41	21	36.2
Kumboedza	0	12	12	100	Middle	42	31.3	24	41.4
Shungu	6	4	10	40	Middle-Rich	22	16.4	9	15.5
Simba Ivhu	5	6	11	54.5	Rich	3	2.2	1	1.7
Survival skills	3	11	14	78.6	Total	134	100	58	100
Total in CA	123	367	490	74.9					
Tagarika	37	13	50	26					
Xanadu A	60	43	103	41.7					
Total in A1	97	56	153	36.6					
Total	220	423	643	65.8					

Source: Compiled by the author based on own survey data, 2018-19

Furthermore, among the female co-operators alone, 41.4% of these were in the middle peasant cluster as compared to 31.3% of males (*Table 7.24*). Approximately 58.6% of the female co-operators were in the middle or upper clusters as compared to 49.9% of men in the same categories. This illustrates how women can take an active role in rural development initiatives, which can serve to propel their material interests.

Although women overwhelm men in terms of numbers, their access to information was highly constrained, and the differences in access to information were statistically significant (*Table 7.25*). That has implications on the female farmer because as highlighted, information is very critical in the agricultural production cycle. Women accessed less information on such issues as 'type of commodities on-demand', 'time that they would be on-demand', and 'supply in different markets'.

Table 7.25: Access to marketing information by gender

	male		female		Pearson Chi-Square Tests		
	No.	%	No.	%	Chi-value	df	Sig.
Commodity prices in different markets	121	90.3	50	86.2	0.696	1	0.404
What commodities are on demand	118	88.1	41	70.7	8.581	1	0.003*
When commodities are demanded	113	84.3	38	65.5	8.529	1	0.003*
Supply in different markets	103	76.9	31	53.4	10.529	1	0.001*
Availability of services eb transport	85	63.9	26	46.4	4.969	1	0.026*

Results are based on nonempty rows and columns in each innermost sub-table.
* The Chi-square statistic is significant at the .05 level.

Source: Compiled by the author based on own survey data, 2018-19

In their 2009 and 2015 survey reports, the SMAIAS (2015) argued that women faced more impediments in purchasing inputs such as fertilisers, weedicides, and certified seeds. My results are consistent with this because lesser proportions of women accessed and utilised variated inputs. Nevertheless, differences between male and female farmers in input access were not statistically significant except for fertiliser.

The same situation was observed within the overall performance of various agricultural production variables. For instance, the amount of income derived from the activities in agriculture for men was US$2021 while that of women was US$2158, meeting attendance rates for men was 60% while that of females was 56%, men hired 5.76 persons of casual labour while women hired 5.88. The same pattern was recorded for permanent labour utilised and the average number of family labour hired in for on-farm activities. The only significant difference observed was that men were receiving 7.5 times more incomes (US$2865) from activities outside of agriculture (from investments) than women (US$385) (see *Table 7.26*). This reality further reinforces my argument that the cooperative offers a way for the women to climb the economic ladder.

Table 7.26: Performance of women and men in the cooperative

	male			female		
	N	% No	Mean	N	% No	Mean
Income from agricultural production in 2014-2017	117	87.3	2020.96	52	89.7	2158.37
Income realise from capital invested outside agriculture	26	19.4	2865.46	13	22.4	384.85
Average savings per households in 2017	134	100	175	58	100	377
Meeting % attendance rate	134	100	60	58	100	56
Casual hired labour average of 2014-2017	68	50.7	5.76	30	51.7	5.88
Average family labour hired in 2014-2017	133	99.3	4	57	98.3	3
Permanent hired labour average of 2014-2017	23	17.2	2	11	19	2
Average amount of family labour sold out of 2014-2017	12	9	1.78	6	10.3	1.61

Source: Compiled by the author based on own survey data, 2018-19

There were no statistical differences in the challenges that men and women faced in selling their cooperative produced goods into the market. Without cooperatives, men tend to dominate women in access to education, information, material resources and market access.

Severity of challenges by peasant classification

Finally, I analysed how different peasant classes were affected by the varying type, nature and character of the challenges that their cooperative faced. The farmers reported that their cooperative was affected by low levels of mutual trust, among members and between members and the management committees. Overall, 65.1% of the farmers regarded this as a challenge, and interestingly, this was

307

highest in the middle-rich (83.9%) followed by the middle cluster (78.8%). These findings perhaps suggest that as the person becomes more uplifted in the social ladder, the more they become suspicious of other members and fear that they are out there to take advantage of them. The latter has far-reaching implications on the cooperative model that can work in rural areas. It will need to come up with a mechanism that reduces areas of mistrust and improve transparency. Issues of trust also affected the Japanese cooperative movement, and thus, a model that can overcome this will appeal globally.

The second most listed problem was the absence of infrastructure in rural areas. In particular, the quality of roads to transport agricultural commodities and people who provide services, modern storage facilities, electricity, and clean water were vital concerns.

Table 7.27: Challenges facing cooperative organisations – by peasant class

	Penury		Poor		Middle		Middle-Rich		Rich		Total	
	No.	%	No.	%	No.	%	No.	%	No.	%	No.	%
Poor mutual trust	6	40	38	50	52	78.8	26	83.9	3	75	125	65.1
Poor infrastructure	7	46.7	31	40.8	44	66.7	26	83.9	4	100	112	58.3
Competitive markets	6	40	25	32.9	45	68.2	29	93.5	3	75	108	56.3
Management & admin	5	33.3	20	26.3	42	63.6	24	77.4	3	75	94	49
Member commitment	8	53.3	31	40.8	31	47	7	22.6	1	25	78	40.6
Value of shares	5	33.3	34	44.7	32	48.5	4	12.9	0	0	75	39.1
Technical skills	6	40	26	34.2	28	42.4	7	22.6	3	75	70	36.5
Bank conditions	8	53.3	26	34.2	30	45.5	3	9.7	0	0	67	34.9
Limited income alternative	2	13.3	15	19.7	25	37.9	18	58.1	2	50	62	32.3
No irrigation systems	7	46.7	27	35.5	20	30.3	4	12.9	1	25	59	30.7
Credit facilities	4	26.7	14	18.4	25	37.9	6	19.4	1	25	50	26
High interest rate	5	33.3	21	27.6	22	33.3	1	3.2	0	0	49	25.5

Source: Compiled by the author based on own survey data, 2018-19

Again, more farmers in the upper echelons of the peasant cluster felt that this was a big challenge. These farmers are no longer satisfied with necessities but are concerned with the more prominent and more rapid development of their cooperatives. Therefore, the way they conceptualized challenges was different from the lower echelons. That is the reason why they also noted their cooperatives as accessing markets for their products, but the markets were not competitive enough as reported by 93.5% of the middle-rich, 68.2% of the middle and 75% of the rich (Table 7.27).

Other challenges that farmers thought their cooperatives were experiencing the most, as reported by between 25% to 40% of the farmers across all clusters were the lower value of shares, lower levels of technical skills, poor conditions of borrowing from the banks, limited income sources and high-interest rates in cases that the cooperative actually found funding. These results will become vital in drawing up a cooperative model in Chapter Eight.

Concluding remarks

This chapter sought out to understand in greater detail the current trajectory of the cooperative movement in Zimbabwe through quantitative data collected in Goromonzi district and from other sources. The main aim was to provide a detailed understanding of situation obtaining in Zimbabwe before applying lessons learnt from Japan and those suggested through Chayanov's theory of cooperatives. Using this data, I carried out a multivariate analysis to describes the different classes that exist in the small-scale peasants as stipulated under Chayanov's theory of peasant cooperatives. I found that the behaviour, production models and challenges of the farmers in the CAs and the A1 were almost similar except in land utilisation, labour self-employment, casual labour use and amount of income from investments outside of agriculture. These four variables distinguished the survey households into five different classes: penury, poor, middle, middle-rich and the rich. Additionally, I found three dominant classes within these five classes, the other two minority classes catered for upward mobility and the other for downward mobility.

Significant differences were observed amongst the farm classes in terms of their access to inputs, access to finance, access to labour (especially casual labour) and differences in their sources of income. There were mixed feelings in the role of the government in the cooperative movements. The majority agreed that the government had too much power; however, there was no agreement on whether this was a good thing or not, and whether the government should maintain or reduce their heavy hand in the movement. I argued that deliberate government policy to limit its involvement to training and capacity building would improve cooperative management.

The Zimbabwean case study highlighted various issues in the management of cooperatives. The results showed that a higher proportion of the nine different cooperatives that I interviewed were women and that cooperatives had managed to improve the lives of their members through improved access to equipment and output markets. Women performed well across the five different peasant clusters and mainly dominated the 'middle' peasants. The group approach, which virtually is what cooperatives are, was being utilised in many instances in the rural areas, for example in extension provision as well as in micro-financing arrangements.

Even though all classes were engaged in asset accumulation, lower-end clusters accumulated smaller assets (hand tools and animal-driven equipment) using mainly proceeds from agriculture. Higher echelons of the peasant classes accumulated agricultural power-driven implements using predominantly proceeds from investments outside of agriculture. I argued that this was a positive aspect because the use of incomes from outside agriculture to buy agricultural assets shows that farmers were willing to do agriculture work (which we want in order to form production-based agricultural cooperatives). Additionally, it showed us that these off-farm sources are utilised because farmers cannot access financial markets for credit and loans. There appeared to be significant differences between the Communal Area (CA) cooperative movement and the new movement in the resettled areas. Although both are getting various forms of assistance from the government, the former appeared to depend a lot on external actors (government and the NGOs) for sustainability, while the latter appeared to be finding its way. Thus, I concluded that there is a new wave of cooperatives on the rise in Goromonzi, which gives scope to refocus efforts towards supporting cooperatives.

The book has so far discussed in greater detail, the agricultural cooperative movements in Zimbabwe and in Japan. Based on Chayanov's theory of peasant cooperatives, I collected and analysed primary and secondary data from villages in both countries. The analysis results revealed a number of similarities, especially in the land reforms and the implementation of respective grain policies. The most significant difference between the two country's post-land reform realities was the robust support of grassroots cooperatives by the state in Japan. The cooperatives helped answer some aspects of the agrarian question in Japan mainly because they amplified farmer's voices within the Community-Market-State relationship. Through the local and national cooperative federations, farmers could lobby the state or negotiate better with the markets to get their concerns addressed. In contrast, this type of farmer voice amplification is missing from the Zimbabwe post-land reform sector. Using lessons from Japan, and an understanding of the current situation of cooperatives in Zimbabwe discussed from Chapter One to Chapter Seven, an alternative cooperative model is proposed in the following chapter.

Chapter Eight

Restructuring Agricultural Cooperatives: A New Cooperative Model for Zimbabwe

> How to combine the community, the market and the state in the total economic system is probably the most important agenda for economists geared toward the reduction of poverty in developing economies, as well as the maintenance of economic vitality and social harmony in developed economies.
>
> – Hayami (2010, p. 118)

Introduction

Throughout this book, we have discussed in great length, developing, and developed cooperative movements in Zimbabwe and Japan, respectively. I obtained a great deal of knowledge and valuable lessons during the process of comparing the two countries after their land reforms: from literature (Chapter Three vs Chapter Four), and from empirical evidence (Chapter Five vs Chapter Six). Several lessons from Japan are presented in this chapter together with how they informed the development of the cooperative model.

Why lessons of Japan are relevant to Zimbabwe.

The ultimate goal of this book project was to develop a cooperative model for Zimbabwe, Chapter Seven provided an in-depth analysis of the Zimbabwe cooperative so as to understand how lessons from Japan can be restructured and incorporated into a new cooperative model (a skill the Japanese knows very well). There are several lessons that Zimbabwe or other developing nations can learn from Japan's agricultural cooperatives. For Zimbabwe, the results of the radical land reform of Japan, and the development of agriculture, bankrolled by an aggressive grain marketing policy is of great relevance to finding solutions for Zimbabwe. Additionally, the study

313

of Japan provided an opportunity to understand a trajectory which cooperatives can potentially take, from establishment, to growth and through to the threats of degeneration. These characteristics are not observable in any developing countries in Africa or the rest of the Global South where cooperatives seem to be stuck in the initial development stages.

However, we have to be sober about the process in which we can directly transplant ideas from one geographical area to the next. There exist significant differences between these countries, which makes it hard to adopt policies from one country to the other (but not impossible). Nevertheless, as one of the lessons, we learnt from Japan; learning from other countries is acceptable and highly encouraged. It was necessary to find ways of modifying or localising these lessons to suit local socio-economic conditions and then adopt them under local terms – this formed lesson number one.

Lessons No. 1; It is OK to learn from others – Throughout the study of JA, we see how Japanese policymakers imported concepts, learned (sometimes copied) from other countries, modified them according to their societal structures and adopted them. For example, as discussed, before adopting the Western model of cooperatives that focused on single-purpose, Japanese scholars modified them to multipurpose. This was done because the countryside required organisations that could provide everything under one roof. The agrarian structure was dominated by small-scale producers that could not afford a membership to two or more different cooperatives (discussed in Chapter Three, page 71 and page 86).

In this chapter, I introduce the readers to a new cooperative model for Zimbabwe. First, a summary of the problems in Zimbabwe's post reform agriculture is presented. Some fundamental assumptions and policy environment that are prerequisites for the model to work then follows. Then the chapter describes the ideal cooperative structure, ideological paradigm shifts, role of the government, the community, the market, and some other supporting institutions required for the cooperative movement to prosper. Throughout the chapter, while discussing the various new ideas,

314

reference is made to how the study of Japan helped in shaping these ideas.

Current challenges in the Zimbabwe cooperative movement

The cooperative model that I seek to develop should recognise and resolve the current and long-term challenges in the agricultural sector and the challenges faced by the farmers within the cooperative movement as well. In order to attain this goal, I summarised the challenges from Chapter Four, Chapter Six and Chapter Seven that are now pulled together into *Table* 8.1. The listed challenges, the lessons from Japan and the theories of Hayami Yujiro and Chayanov drove the formulation of the new agricultural cooperative model described henceforth.

Table 8.1: Summary list of challenges observed in Goromonzi cooperatives.

Challenge (most affected: S = State; C = Cooperative; B = Both)		Brief Description of Challenge
1.	Unregistered cooperatives (S)	There were too many unregistered cooperatives, especially in the agricultural sector, because registration is done in Harare. Unregistered cooperatives complicate cooperative law administration.
2.	Limited budget allocations (S)	A handful of staff in the Department of Cooperatives had no time for field visits because of inadequate resources.
3.	Outdated legislature (B)	The Cooperative Societies Act of 1996 is outdated and too broad (covers all sectors). It now has many loopholes.
4.	Unconducive macro-economic environment (B)	ESAP, inflation and dollarization had a significant negative influence on the movement as people lost their savings. People lost confidence in cooperatives and SACCOs fearing loss of their hard-earned money. Corruption and politics also undermine cooperative development.

315

5.	Frequent Changes in parent ministries (C)	Each time this happened, it set the movement a couple of steps back as information was lost with each change.
6.	Anti-Cooperative state (C)	Government initiative has mostly been dormant over the past 25 years, especially after the land reform program.
7.	Politics (C)	Most of the cooperatives on the ground are too political for the benefit of politicians and not members.
8.	Low debt repayment (C)	A challenge for the success of SACCOs is the low re-payment rates from borrowers (farmers), only 15-30% repayment rates.
9.	Education and information (C)	Farmers need to understand the concept of cooperativism first because many believe it is a means for the state to give free inputs to farmers
10.	Ideology (C)	Farmers' attitude towards the cooperative is a big problem. Thus, each time they get a loan, they are not obliged or do not feel the need to repay it. Donor and government free inputs lead to donor syndrome, and hence ideology needs to be stronger to counter this problem. Farmers do not follow any ideology, no focus on one crop; people are not working together to achieve development goals.
11.	Mismanagement (C)	Dishonesty is rampant. A lot of the chairmen were hiding their receipt books from their secretaries and their management committee.
12.	Trust (C)	Levels of trust were low, between members and the management, and within the management itself.
13.	Government hegemony (C)	The state does not acknowledge the existence of information asymmetries between them and the farmers. Their all-knowing stance is counter-productive.
14.	Compromised apex body (C)	Zimbabwe National Federation of Cooperatives (ZNFC) is incorrectly constituted, hence, compromising its independence from the State.

Source: Created by author, own study

Prerequisites for the Zimbabwe Community-Market-State system

In Zimbabwe, cooperatives have a tainted history or image such that the mention of the cooperative model in development studies takes people back to socialist government policies and the failures of the collective farms in the 1980s. This image requires redress, and more precisely, it needs to be exorcised before genuine cooperatives are reinstalled to their rightful place in development. I present three major areas that stand as prerequisites and the tentative ways in which the government, the peasants and the markets can try to secure them.

The role of government and policy instrument
There are two major lessons we get from the Japanese system and also from theory on the relationship of the state and the cooperative movement:

Lesson No. 2; Cooperative requires the government – The structure of the MAFF revealed a more sober governmental approach to cooperative development because there was a dedicated division that oversaw agricultural cooperative business. Government funding is still vital in support of cooperatives and farmers. Countries like Zimbabwe that have one harmonised cooperative societies law, which is administered by a department in the Ministry of Women Affairs have a long way to go in this respect (discussed on page 94).

Lesson No. 3; Agricultural cooperatives can strengthen the peasant path – And from a Chayanovian perspective, agricultural cooperatives can strengthen the peasant path (discussed on page 52) which, as identified by several scholars and multi-lateral development practitioners, is the best path for Africa's development. Additionally, the cooperative and the government complimented each other in the ultimate quest to develop the rural areas in Japan. In this way, the cooperative system made it easier for the state to formulate and implement rural development policies (discussed on page 70 and page 99).

The gap between government and farmer perception of rural problems complicates the CMS relationship – information asymmetries as Hayami Yujiro (2005; 2010) called them (see also Stiglitz, 2002, p. 469). The primary aim of the government and its policy instrument is to ensure a legal basis for sustainable existence of cooperatives. As the analysis proved, although the current policy provides the legal basis for cooperative existence, other fundamental issues such as sustainability, autonomy and democratic decision making are not protected. The Act gives too much power to the state, which undermines the agency of farmers in CMS framework. The government of Zimbabwe should smoothen the functionality of cooperatives by improving the economic environment, policy instrument, political environment, state-level management, and financing.

Economic environment

The majority of problems that are faced by the cooperatives are linked to the unconducive economic environment that it operates. Fixing this is one of the biggest hurdles in the sense that cooperatives should actively help stabilise the economy by improving production and rural incomes. Yet, cooperative development also requires a functional economic environment. The economic environment can be improved to facilitate cooperative growth by ensuring the operation of some of the fundamental institutions. For example, the country has struggled to produce a functional agricultural policy since 2000; and there is low confidence in the current money system in the economy. People in rural areas may not easily access the conventional methods of payment, such as mobile payments (Ecocash). Zimbabwe needs to have a currency that is based on human, infrastructure, and natural resources (see Chapter Six, page 240).

Policy instrument

Here lies one of the biggest problems. As highlighted earlier, the state first needs to establish a separate agricultural cooperative law to overcome several challenges specific to the sector. Secondly, the state should then amend or formulate a new agricultural cooperative law that dutifully adheres to the ICA (1996/2001) cooperative principles and recognise the complexities of agriculture and the consequences of the FTLRP of 2000. These two processes are imperative and urgent. Furthermore, the new agricultural cooperative law should pay attention to i) optimise the functions of the minister/registrar, ii) administer the law under the Ministry of Agriculture, iii) ensure legal actions for unavailability of the cooperative register (contravention of the Act), iv) decentralize registration of cooperatives, v) specify minimum requirements for committee members to Certified Farmers, vi) remove the hegemony of the government in the Federation, vii) and also establish a central cooperative fund without fail (see Chapter Six, on page 235).

Political environment

The rural areas are one of the most active areas of mobilisation for the ruling party ZANU PF, and thus, policies or programs tend to be coloured by party politics. While it may be difficult, there is a need for considerable efforts to depoliticise agricultural development, especially when it is misused to cover corruption and abuse of privilege or power. It makes sense for any party in power to want to retain power at all cost; however, I argue that it is time that ruling powers ensure that they stay in power by making the people happy through development. The cooperative should be owned by the people so that they can succeed by themselves (not owned by the government or NGOs). Government influence is supposed to be limited so that farmers can influence government policy formulation processes when they are within the cooperative and not as individuals (discussed in Chapter Six, page 225).

Management of cooperatives

In addition to reducing its influence in the National Federation, the state needs to improve its management of cooperatives. The first task is to transform the administration of the sector-specific cooperative to its respective ministry. For example, agricultural cooperatives should be managed by the Ministry of Agriculture, while mining cooperative should be the prerogative of the Ministry of Mines and Mining Development. Each ministry is better equipped to deal with sector-specific issues. Also, upon the change in government, no information is lost as had happened in the past when the administration of the Act changed ministries. Then a comprehensive cooperative register should be developed to curb corruption issues and abuse of cooperative benefits through fake cooperatives. Interview data revealed confusion and power struggles within the Department of Cooperative Development. These issues need to be sorted with immediate effect.

Furthermore, the government is supposed to engage with the farmers and involve them in the formulation of policies and their eventual implementation. While the state may argue that all their new programs had farmer participation, the situation on the ground suggests high levels of hubris. Quite crucially, they need to give the cooperative movement some breathing space, starting with clarifying conditions for running and election of chairpersons and officers of the NFC. The federation should have a representative from each of the cooperative economic sectors. The government should hand over some of the critical functions that it does to the NFC to reduce cost and also to make it more efficient. The government can focus on other issues if they 'delegate' this task to the national movement (see Chapter Five, page 169).

State Subsidies and Financing

Historically, the state mainly financed the cooperatives in Zimbabwe, and this has often undermined the autonomy of cooperatives. The state needs to re-evaluate the focus of its financing to ensure that they do not perpetuate the donor syndrome, which undermines cooperative development. The subsidies debate has stood for a long time. While I do not dispute the need for funding of agriculture, the methods and channels of the funding require restructuring. A focus on the funding of agricultural training and skills development will go a long way into resolving this challenge. I argue for the state to stop the system of free inputs to individual farmers and instead to provide for the establishment of a cooperative fund and capitalise it. Then the funds can be accessed by the farmers through the cooperative (or otherwise) at concessionary rates. Farmers need friendly and conducive payment plans and not free inputs. This move will work as a double-edged sword by answering parts of the agrarian question of finance and extinguish the dependency syndrome that affects agricultural producers (discussed on page 240).

Finally, the government should be willing to learn from other countries such as Kenya and Rwanda, where cooperatives are changing people's lives through SACCOs. There is a need to embrace SACCOs at the government, national and local levels as well as to carry out extensive SACCO training workshops across Zimbabwe. The government has only recently started efforts to establish SACCOS and have developed plans for this institution for every type of cooperative. Other international organisations (ILO, UNDP, Swedish Cooperatives and World Health Organisation) are willing to help Zimbabwe achieve this. The model might work given the fact that the greatest challenge to the Zimbabwean farmer is the lack of finance for production.

Education, Skills & agricultural training

Agricultural education, skills and training has primarily been driven through government policy and provides one of the best ways to improve the nature, character, and quality of the peasantry. I argue that the government should have embarked on extensive agricultural skills training exercise before the implementation of the FTLRP to ensure that land beneficiaries were in a better position to produce. The same should have been done before the 1980s resettlement programs and before putting farmers in collectives. If it had trained farmers and government officials, the policies would have yielded desired objectives. In addition to giving farmers knowledge about procuring finance, inputs, production and marketing of their crops, extensive agricultural training would have given the government a chance to reorient beneficiary ideology (in terms of explaining the objectives of the reform, farmer's role versus that of the government and how they all could partner with the private sector). These issues seem to be missing from the current CMS relations with many believing that just as they got the land is the same way that they should get finance and inputs necessary for production.

Institutions of agricultural training should be able to address some of the challenges in farmer attitudes, involvement in politics and corruption as well as improve management practices. I expect the government to intensify its effort in the form of funding for education and training. I propose the establishment of a mandatory course in agricultural training which can be instituted by most of the agricultural training institutes already in existence in Zimbabwe. With continued yearly distribution of subsidies and free inputs to farmers without education, skills and training on how to adequately use them, agricultural production will remain low.

The role of the community (the peasantry)

Lesson No. 4; Unified farmers are hard to exploit – The JA has been under pressure to degenerate over the past 30 years; however, farmer agency remains strong because of the cooperative system. The TPP question stands as the most significant evidence of continued farmer agency in recent times. If a population of 2.1 million cooperative members (out of a total of 128 million people in Japan) can amass such power through the cooperative, the possibilities for the agrarian population – which comprise almost 70% of total African population – are unimaginable. Another example is seen through dictating reform of agriculture by the JA, resulting in what is known as JA self-reform. In 2015, the cooperative managed to strike a deal that allowed JA-Zenchu to convert to a corporate association but maintained the status quo of JA-Zennoh and its associate members. This highlights the strength of the agency of farmers secured through the cooperative model (discussed on page 82 and page 104).

The success of the cooperative movement henceforth will depend significantly on the nature of the peasants. I argued that the resultant agrarian structure was ideal for the development of the cooperatives, especially in the newly resettled areas where cooperatives were formed for and by the farmers themselves. However, more needs to be done in terms of inducing a significant paradigm shift in management practices; skills, education & training; and socio-economic behaviour (attitudes, politics & corruption) during agricultural production. In light of the pervasiveness of information asymmetries that rendered state and market interventions inadequate for rural development, Hayami (2005)

322

strongly argued for a role of community that is expected to provide public goods (including fair institutions and even infrastructure). Thus, the community is expected to supplement government efforts to address market failure (undersupply of public goods) by filling in the information gap because only they know more about themselves than the state and the market (Stiglitz, 2002, p. 471).

Socio-economic behaviour (attitudes, politics & corruption)

Based on Hayami (2005), the community needs to foster bonds of communication and trust within the community as this has ripple effects throughout the cooperative system and the interaction with the state and the markets. While cultural, family and tribal bonds are active in the CA, the same is yet to hold for resettled areas and hence the newly resettled farmers need to foster improved trust and communication through other means such as a church, sports, field day events or other recreational activities (village or location-based) (Mkodzongi, 2016, pp. 75–78). Trust and communication are critical because it enhances mutual observance of traditional norms and conventions within the peasantry, thereby reducing transaction costs.

Since many problems came from the way farmers viewed cooperative programs, farmers should change their attitude. Chairmen and directors of cooperatives need to be honest when doing their duties as this has a profound effect on the smooth operation of the cooperatives. There needs to be a program that shifts people's mind from their current way of thinking, then reduces levels of ignorance within management committees and also among the members through intensive education and training programs. In the pre-independence movement, white farmers firstly had to understand cooperativism through theory and practical perspectives. This is what is lacking in the movement that came after independence and the ones that are operating today. In addition to proper education and ideological stability, a system of rewards for good behaviour in order to encourage more and more people to practice honest business should be developed. More commitment should also be shown by the cooperative leaders and from the rest of the management team. It is indispensable to improve the level of trust and communication within the movement.

This study revealed the difficulties for farmers to stay away from politics, but it might not be beneficial for the cooperative to be ultimately apolitical. In fact, most of the cooperatives on the ground are too political. Either they were formed by the state (top-down approach) or the management committee deliberately engaged themselves with politics in the hope of getting favour from the political elites. I argue that engagement with political apparatus should be avoided and that in any case, cooperative decisions should always be democratic and should benefit the farmers more than they do to the politicians. The government of E.D Mnangagwa had managed to remove some of the power that some corrupt chairmen had but only those aligned to former president RG Mugabe. It is hoped that new political machinery will not create new corrupt chairpersons aligned to the new dispensation (see discussion on page 193 and page 237).

Management practices

There is a need to improve management skills and practices which mainly depend on the two levels mentioned above. One of the most significant challenges in the cooperative is debt and subscription fee collection, which resulted in as much as more than 50% of cooperative revenue loss. One way of resolving this (as suggested in an FGD) is the use of incentives such as early-bird incentives. Farmers who pay their loans earlier are given a reduction such as a 10% discount on their total loan if they pay before a specific date (subject to negotiations with the lending bank).

There was a need to tighten the input distribution mechanism in order to reduce both transportation costs as well as time taken to get the inputs. Although people who do not repay their loans are counter-productive to the movement, chairpersons and directors still need to engage with them with respect and high levels of diplomacy since the fundamental requirements of cooperatives is cooperation, peace, and inclusive development. Engagement with local and traditional authority were other ways of drumming support for the chairmen to do their jobs without fear. However, chairpersons should exercise precaution since the local authority is also riddled with politics (on page 182 and page 216).

The role and nature of agricultural markets

Lesson No. 5; Cooperatives can perform in liberal markets
– The book revealed how the development of capitalist Japan affected the cooperative movement and in return, how the movement managed to survive and become one of the most profitable businesses in Japan. This revelation serves as a big lesson for developing countries that seek to integrate and modernise rural areas into global markets. The JA collaborated with local and international companies along business and market terms. The institutional structure of cooperatives enables establishment of cooperative corporations as seen in Japan. In the long run, these cooperative corporations can take over the liberalisation of the agricultural sector (discussed on pages 70, 94, 99 and 177).

Agricultural markets are fundamental in the CMS framework because these are platforms for coordination of profit-seeking individuals based on supply, demand, and price restrictions. In summary, the CMS framework highlights that the state (through coercive power), the community (through consent-based coordination) and the market (through competition informed by supply and demand) are the three critical prerequisites for development (Hayami & Godo, 2005, p. 311; Otsuka & Kalirajan, 2010, p. 224). In most developing countries, including Zimbabwe, the state and the market are present and exercise power over a vulnerable community. Cooperatives will help strengthen this third player. The study expects the markets to be a source of capital, technical know-how and most importantly of technology. Modern agricultural production will depend on advancement in technology adoption and use, which depends on how the community interacts with the market. Markets will significantly improve with cooperative participation (see discussion on page 26).

Critical assumptions about the Zimbabwe peasantry

Lesson No. 6; We are different but the same – This book reviewed many similarities in rationale, implementation, and challenges faced during or after the land reform in both Japan and Zimbabwe. The reform is a crucial point of comparison. The lesson here is that despite temporal and spatial differences, the nature of motivation, challenges and some aspects of the execution of the land reform was virtually the same. The differences only lie in the advancement of one country along the same trajectory. This is the essence of leap-frog development strategy. Hence, a study of specific periods along the trajectory can be beneficial to the country that is yet to pass through the same stage. For example, in Japan, the constitution and the agricultural policy regulates the sale of agricultural land to prevent the return of landlords or reverse the 1945 land reform. The agricultural basic law forbids the sale of land on the open market, and sales are made only through the state. While the rest of the market is open, the market for agricultural land is not. The Zimbabwe Land Administration has been battling to come up with a similar law. Lastly, the rice control act closely resembles the 1980s-1990s maize marketing policy of Zimbabwe (see Chapter Three and Four), and these similarities strengthen the rationale for a comparison which we carried out (discussed on page 78). This line of thinking helped us to rethink about the type of peasant to be targeted by the new model.

The definition of the peasantry has been debated in literature as I discussed in Chapter Two. The nature and character of the peasants (encompassed in the definition) has a profound effect on the cooperative that we can encourage in rural areas. In order to provide an alternative model, I made the following assumptions about the peasantry, based on literature, empirical data, and field observations.

A. There are three dominant classes in the cooperative movement: (mainly) peasants, a middle and an upper class.

B. The peasantry seldom employs labour outside of family labour; when it rarely does, it is a casual type of labour.

C. Although farmers are primarily concerned with achieving a minimum level of agricultural production necessary for social reproduction, they cannot escape the currency and commodity circulation (the markets) and hence offload excess onto it.

D. The small farmers have expanded access to land in light of the proliferation of land renting and leasing markets after the FTLRP (though not yet legalised).

E. There is a minimum level of education necessary for skills/agricultural training. There are NO statistically significant differences in the level of skills/education across households or classes.

F. There is capital accumulation from agricultural income.

The assumptions above simplify the formulation of the cooperative model.

A new structure for the movement

The current cooperative structure has eight sectors; each sector has a primary level, secondary level and an apex level (see Chapter Four, page 144). The overall association that unites all is the Zimbabwe National Federation of Cooperatives (ZNFC). The general message I seek to push with this new model is that nothing is for free, and every new cooperative need to work hard to achieve development goals. This model proposes tangible milestones that can help a cooperative grow internally and also develop from a broader perspective.

Membership structure
Lesson No. 7; Cooperative freedom-equality trade-off – The Japanese cooperative has more equality and less freedom than Western styles. The question, then is what is the most ideal, which one should we adapt and modify in the case of Zimbabwe? How can we reimagine the British-Indian cooperatives that exist already so that they achieve ideal freedom to equality ratio? (discussed on page 99).

The first issue I would like to discuss is that of ownership rights of cooperatives. In this model, all members are equal in terms of their importance, but the focus will be on full members and commodity producers. These were topical in the Japanese cooperative movement, but for the case of Zimbabwe, where I proposed that cooperatives should begin as small single-purpose cooperatives, the issues may not be problematic (see page 329). However, these will become apparent in the long run, and hence an idea of how to handle them is discussed in this sub-section.

One of the most significant issues with membership is getting sufficient numbers of people to join the cooperative. Thus, there is a need to increase the incentive for people to become full members. In most cooperatives in Zimbabwe, there is the exclusive use of cooperative facilities, and in some cases, non-members cannot trespass on the cooperative property. While this makes sense, it does little in terms of advertising and attracting people to the cooperative. There needs to be more contact between the cooperative and the non-members to encourage them to join, become associate members and eventually full members.

The current cooperative structure has full and associate members. The difference between these two is primarily the voting rights and the ability to be a candidate for the committees or director/manager. Data revealed that some members were interested in using some of the cooperative facilities but were not interested in joining the cooperative or having the burden to pay subscription fees, attend meetings and become leaders. To cater to these types of farmers, and based on insights I learned from Sanbu Yasai Network of Japan, I divide associate membership into two types. I would like to propose three types of membership; 1) Full members, 2) Associate member tier 1, and 3) Associate member tier 2 (see Table 8.2). The Associate member tier 1 is the usual associate members. Associate member tier 2 is an associate member that can attend meetings and express their opinion/idea but cannot vote or stand as a leader; they do not pay subscriptions but only pay the cooperative a commission when they sell their products through the cooperative (for example, 5% commission).

The associate member has restricted access to cooperative facilities, cannot benefit from price discounts or member subsidies and cannot earn dividends when the cooperative makes a profit. The main aim is to encourage more partnerships with non-members so that they can understand how the cooperative works and why they should join it.

Table 8.2: New three-tier membership structure

Aspect	Membership status		
	Full	Associate member	
		Tier 1	Tier 2
Attend meetings and express opinions	Yes	Yes	Yes
Subscription payments	Yes	Yes	No (only commission)
Use of facilities	Unlimited	Unlimited	Limited
Cooperative education	Free	Free	Pay a fee
Dividends	Yes	Yes	No
Subsidies	Yes	Yes	No
Price discounts	Yes	Yes	No
Voting rights	Yes	No	No
Leadership candidature	Yes	No	No

Source: Created by the author based on own study, 2019

Geographical structure and stages of development

Lesson No. 8; Long-term planning and consistency – We also see the need for a broader vision and national long-term cooperatives plan which should not depend on the too volatile government policy (this should also be a prerequisite). There should be Three, Five, Ten, twenty-five and fifty year plans going into the future. The story of Yotsuba in from page 158 shows the strength of long-term plans, goals and visions while the stories of Xanadu A. and Tagarika (who had no idea about the future of the tractors they had received – page 243) shows the lack of long-term planning.

The current cooperative structure is federated, which means that local members belong to local level cooperatives which then belong to the secondary cooperative. These cooperatives can then belong to the central association. No individual member can belong to the central association unless through the secondary level cooperative. I propose to maintain this structure given the fact that there is a three-tier cooperative structure, and I advocated for many smaller cooperatives at the primary level in the beginning.

As the cooperative movement develops, I argue that it should move towards mixed governance and control structures (*Figure 8.1*). A mixed structure is flexible and when individual cooperatives have grown into multi-purpose, belonging to a single-purpose provincial level cooperative may not make sense. Hence a shift to mixed structure enables the primary multi-purpose cooperative to become direct members of the Association. I also believe the current one village one cooperative structure should be maintained considering difficulties in communication if members join cooperative in different jurisdictions.

Figure 8.1: Gradual development from a federated to a mixed cooperative structure

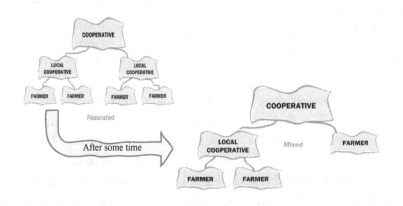

Source: Created by author, based on USDA (2011, p. 2)

Significant structural changes to the current cooperative structure

In coming up with a new cooperative structure, I paid attention to the following key areas that require restructuring:

1.   The Cooperative Societies Act of 1996 should be dismantled into different sector-specific cooperative acts. A method of doing this should be gradual and over five years or more years for the government and the stakeholders to gradually factor the changes into their institutional programs. I propose firstly dismantling the SACCOs, Agricultural Cooperatives and the Housing Cooperatives into different laws that can be administered by the sectorial ministries. The individual acts will be administered in their respective ministries, i.e. the Agricultural Cooperative Act should go to the Ministry of

Agriculture. There should be minimal changes in the parent ministry after that.

2. The movement needs a cooperative bank or a financing institution which can take over the handling of most agricultural financing. This bank should make use of the cooperative certificates and farmer licence criterion. However, this process requires time and a strong financial base, and thus in the short run, partnerships can be forged between the cooperative movement and the AgriBank of Zimbabwe to play this role. Another avenue is tentatively to have the SACCOs play the role of a cooperative bank.

3. I believe that more emphasis should be put on the primary cooperatives, which are genuinely farmer-driven. The research has proved that the probability of farmers forming a cooperative increased if farmers felt that they could benefit from being members. Thus, I recommend that more rewards should be on farmers working through cooperatives (for example a 2% price incentive if selling through a cooperative, or better chances of getting credit if applied through a cooperative).

4. There should be a Zimbabwe Good Agricultural Practices (ZiGAP) institution at the national level, which can oversee the certification of farmers as well as monitor safe agricultural standards and quality. The ZiGAP can work hand in hand with other economic sectors and will help build trust within the farmers, as well as with the private sector. Ultimately this will lead to the building of trust of Zimbabwe agricultural products.

Development of the cooperatives

I propose a gradual transition from the current structure, beginning with the amendment of the act (to separate the sectors). The World Bank report (1989) recommended sufficient amalgamation of small primary cooperatives to become more prominent cooperatives of at least 500 members in order to achieve economies of scale. Although I agree with this in the long run, in the short-run (5 to 10 years), I propose for Zimbabwe to take a gradual increase in membership starting from small cooperatives. Because of low levels of cooperative education and understanding of cooperative principles and ideology, cooperatives should start as small single-purpose organisations for at least five years in which other potential members can learn and then become members.

Figure 8.2: Transition from Single to Multipurpose cooperatives

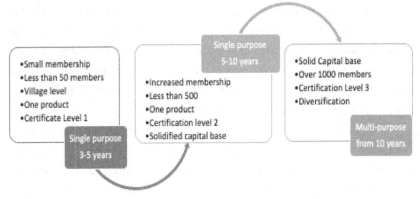

Source: Created by the author based on own study, 2019

Upon reaching a certain point (measured by the cooperative certification score and the member's levels of farmer licence scores in the ZiGAP), a cooperative can become a multi-purpose one focusing on various other businesses (*Figure 8.2*). Applications should be sent to CACU (with ZiGAP certification) which then carries further investigations to see if the cooperative has achieved the milestones necessary for upgrading. Upon ascertaining this, CACU will forward the application together with its recommendation to the ZNFC. The federation can carry simple background checks, if satisfied, then send to the Ministry (with a second recommendation) for registration (the ministry can further inquire, but not necessarily a requirement). This process maintains standards and reduces corruption.

I should further clarify that the cooperative I seek to establish is not only marketing, purchasing or services cooperatives but also focuses on agricultural production. In a similar vein, a gradual increase in the cooperative portfolio, cooperatives should start off as credit cooperatives as suggested by Chayanov. Eventually, as their financial capacity allows, they can broaden into marketing, purchasing, producer and service provision.

Purchasing, marketing, distribution and service strategy

Lesson No. 9; Cooperative identity and cooperative companies – This concept begins at the village level where JA supported local cooperative to focus on the production of a specific crop. In this way, we would know and identify quality with a specific region, for example, oranges from Mazowe in Mashonaland East, or sugar from Chiredzi, or bananas and avocados from Chipinge farmers. The farmer/cooperative certification model can support cooperative identity. We also learn of the use and creation of cooperative companies in the long run which may aid in the attainment of a robust corporate identity. The cooperatives wholly own these companies.

The cooperative should ensure fair trade in marketing (all involved parties must benefit from the trade of a product) and develop distribution channels directly to the market/consumer. The focus should begin on the reduction of transportation cost through bargaining, buying in bulk, buying earlier and accessing input markets otherwise unavailable. This part of the new cooperative model should take centre stage because it helps the members to control and own the product. Eventually, as the cooperative becomes prominent, complex processes such as grading, value addition, and research can be integrated.

I learned the importance of long-term planning in the case of the Japanese cooperative, which had five, ten and twenty-year planning strategies to guide their current decision-making processes. In Zimbabwe, purchasing and marketing cooperatives do not have long term plans. I propose that cooperatives should have laid out plans for at least three proceeding seasons for each cooperative member to help plan purchasing and marketing schedules. Some cooperatives could not enjoy lower prices that come with purchasing inputs six months before the start of the season. Also, planning would help secure markets before the production season kicks off. The cooperative certification would ensure farmers stick to this system.

Using the Japanese method of learning from others and modifying before adopting, I propose a deliberate strategy of cooperative identity to be employed. The concept of one village one product was successful in Japan rural areas, and I would like to modify and adapt it to Zimbabwe. Given that different geographical weather conditions in Zimbabwe, cooperatives can adopt crop-specific branding and marketing strategies. Goromonzi, for example, produces good quality tomatoes, and hence, tomatoes be marketed as a Goromonzi brand. Under the non-cooperative strategy, this is not possible as individual farmers do not have the capacity or cannot maintain standardised products. That is the advantage of a cooperative. Other areas can adopt the same strategy, such as the wild plums from Mutoko or Mopane worms from Beitbridge.

Management strategy and incentives
Lesson No. 10; A certified farmer is an influential farmer –
The new Japanese Agricultural Cooperative Law of 2015 stipulates that the new leaders of the cooperative should be certified core farmers (and should be most qualified). Thus, farmer certification is essential as a gateway to many privileges, including getting loans easily. The certificate is also used when buying Japanese farmland (to get a farmer certificate, one needs to be involved in farming or agricultural production) (discussed on page 165 and page 191). An institution called the Japan GAP Foundation does farmer certification. Farmer certification is a critical lesson for Zimbabwe, especially concerning the establishment of an autonomous body responsible for assessing farmers, training, educating them and then conferring them with farmer certificates. Zimbabwe, which already has some farmer certification, could learn how to make more use of these farmer certifications in the future. The certificate can be a powerful tool used in acquiring loans, credit, and selling to markets. It can be used to monitor and evaluate good behaviour and excellent management skills of farmers. It should reward those that score higher to incentivise those that score lower. A certificate that can be recognised by banks and financial institutions and should have several other non-monetary special privileges. The concept can also be expanded to cooperative certification. Using their certificate, a group of farmers in a cooperative can apply for bank loans just in the same way that a group of scholars with recognised academic prowess can team up to form a consultancy team and get research grants.

Cooperative movements across the world have management issues. One of the best ways to encourage proper management is by choosing good leaders. Good leaders can only be chosen by knowledgeable members that know what to look for in a leader. In this respect, I reiterate emphasis on education and training of the cooperative members as well as of the leaders. The system of farmer licences I propose will go a long way in resolving this problem. Again, my model assumes the government and the national federation would begin extensive farmer education such that strict adherence to such leadership prerequisites can come into effect in the next four to five years. In the beginning, the rules may be relaxed as people get training. Farmer education is critical to the success of not only this model but of agriculture as a whole. Farmer education will improve levels of understanding of cooperative ideologies and principles.

Additionally, as my analysis proved, stronger ideological foundation coupled with open or free flow of information enables easier management of the cooperative. There is a need to encourage both members and cooperative leaders to always act in good faith to avoid corruption and abuse of cooperative assets. I propose a system of rewards for good behaviour in order to encourage more people to practice honest business. The problem of bad debtors, for example, people who can manage to repay their loans to the cooperative should be rewarded through a discount as an incentive. Even small incentives such as T-shirts or caps, farmers will be delighted with such a gesture.

Lastly, although it is challenging to avoid local politics and that indeed the cooperative cannot afford to be politically neutral, I propose that the management should avoid political situations as much as possible. Political consciousness is essential, but as the cooperative movement is still small, it cannot afford to take too strong a political stand.

Improving the ZNFC and CACU

As I discussed in great length in the theory chapter, a robust and functional National Federation of Cooperatives is extremely necessary. Interview data revealed a federation that was unrepresentative of the economic sectors, and the methods of choosing its leaders and directors amounted to state-meddling in the affairs of the cooperative. Thus, I would like to propose strict adherence to the law by the state in terms of Part XI, Section 89 (1)(b) of the Zimbabwe Cooperative Societies Act of 1996 which gives powers to the societies to appoint their preferred chairman and other officers. The ZNFC needs to be active together with the CACU, especially in sourcing funding from the government, the market and other multilateral stakeholders to finance education, training and skills development.

I argue that the Central Association of Cooperative Unions is weak and has been weak for too long a time. For example, in the 1980s, it employed less than five people at its main offices, a situation that still prevails at the moment with less than ten people working there. CACU requires a structural change in which it becomes more prominent by taking over some roles previously done by the government as well as the secondary level tier (to give them more space to deal with an expanded primary cooperative). Thus, CACU needs more financial resources such that it takes-over training, education and skills development of the cooperative members in the short-run (5-10 years). In the long run, this body should oversee the importation of bulky inputs directly and the advancement in cooperative planning, research and information dissemination. As of now, the last cooperative research publication obtainable from the CACU library was done in 1988 while no dedicated website exists for the Association. CACU can also be one of the firms responsible for auditing primary cooperatives.

I propose to do cooperative registration through three different institutions, and these should tally. Each of ZNFC, the minister (registrar) and the Association (CACU) should be involved in the registration and should keep a copy of the register. This process will reduce instances of corruption while improving accountability and access to the information within the movement. A strong CACU will strengthen daily cooperative management while a powerful ZNFC will unify single cooperative voices to protect the interests of farmers (especially in national politics).

Central Cooperative Bank (CCB) and Central Cooperative Fund (CCF)

Lesson No. 11; Agricultural financing – This was key to the development and sustainability of the cooperative movement. In Japan, it was done by the government, mainly because the government is financially resourceful. However, for the Zimbabwean cooperatives, there is a need to set up something similar to the FILP of Japan financed through a combination of government and private sector. Additionally, cooperatives should be self-funded through the establishment of a cooperative bank such as the ACC to handle the farmer's money and be prepared to give them loans (discussed on page 97).

The current cooperative Act (Part XII, Section 91) provides for the establishment of a CCF in which cooperatives should contribute 5% of their net earnings (and any other sources including from other countries) in order to fund such requirements as education and training. However, donations from other countries are made through the Minister. Given the fact that most cooperatives are under-resourced at the current juncture, I propose an active fundraising scheme by the ZNFC from multi-lateral donors, other prominent international cooperative movements (e.g. the JA of Japan) and the ICA. This money should be used strictly for education and training of farmers.

While the prospects of a CCB at the moment may be far-fetched, I argue that it is necessary for the long run. Private financial institutions in Zimbabwe have neglected agriculture food crops in the past 20 years (Mazwi et al., 2019). However, if a CCB had been active, it would have kept productivity high when the financial markets had neglected the newly resettled farmers. The CCB would reduce the emphasis on the provision of collateral which the private financial sector has been adamant about since the inception of the FTLRP. As I argued, farmers bring in a considerable amount of foreign currency and have a right to access finance for this reason. While we develop the cooperative movement, it may be beneficial for the movement to discuss partnerships with the government-run AgriBank so that it may provisionally take the roles of a CCB. In addition to the AgriBank, proper constitution of the SACCOs has

337

the potential to help finance and provide a basis for the formation of a CCB in the long run. Thus, I propose the AgriBank and the SACCOs to spearhead agricultural cooperative financing in the short run. Interview data highlighted that the government was already focusing more on SACCOs as an agricultural cooperative financing mechanism (Mada interview, March 2018).

The overall structure of the Zimbabwe agricultural cooperative system

The overall structure of the cooperative, therefore, should have three types of cooperative members at the bottom to encourage more extensive use of the cooperative through attracting non-members to have a 'taste' of cooperative benefits. The local, district, city and town cooperative would make the first tier and consist of different types of single-purpose cooperatives in the short run, but with the aim to transform into multipurpose upon reaching specific goals in the long run. This process should be the aim of the cooperative as a way of encouraging them to work hard towards cooperative development.

The second tier is necessary because the primary cooperatives will be smaller and numerous. The third tier will be the CACU which should be responsible for national-level marketing, purchasing, services and creation & maintenance of a cooperative brand (*Figure 8.3*). The federated and three-tier should be maintained in the short-run, which should change to a mixed and two-tier structure in the long-run (say after ten years). This process should be the plan, again as a way of encouraging development at the national level. The ZiGAP will provide certification services, encourage more significant information flow, build trust as well as the Zimbabwe agricultural brand at local, provincial and national levels. Attainment of a cooperative certification should be held in esteem.

As explained, the CCB and AgriBank can be the financial institutions of choice at both provincial and national levels. These (together with provincial level, SACCOs and local level cooperatives) will get supervision, training and education from the ZNFC through CCF funding.

Figure 8.3: New Zimbabwe Cooperative Structure

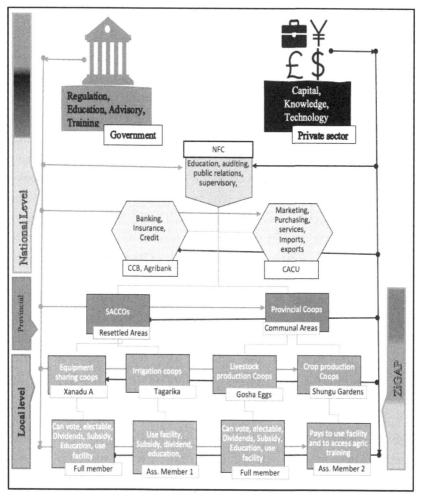

Source: Created by the author based on own study, 2019

In *Figure 8.3*, I highlight the importance of the government and the private sector. The state should reduce its role to just regulation and support of the movement through administering of a revised cooperative law, through education, training and skills development.

The state should minimise the provision of free inputs and machinery and focus on education and training subsidies instead. On the other hand, provision of technology, technical know-how and capital should be the role of the private sector. The private sector is essential and will not directly exploit the farmers in this structure, but will interact with them on an open platform.

Certification/licencing of farmers and cooperatives

I have discussed how the Japanese emphasize the use of farmer certification in land markets, in credit markets, in inputs lending and even in output markets. I would like to modify and improve this concept to suit Zimbabwe conditions. The Zimbabwe agricultural sector has high transaction costs which include the cost of acquiring market information. Information asymmetries in this respect have continued to hamper government and private financial sector dialogue (e.g. use of current tenure documents as collateral). The scoring of financial behaviour widely used in the USA (credit scores), or Japan (farmer certification); have also been proposed for Kenyan farmers (credit scores) in a study by Nyafwa (2019). However, in each of the above cases, it was not used to promote cooperative development. I seek to establish the scoring index for Zimbabwe agricultural cooperatives where scores do not rely solely on private property collateral. In order to overcome some of these challenges, my research proposes the establishment of a national farmer certification/licensing program which would be responsible for assessing farmer capabilities and productive capacity.

The scoring index will go beyond possession of collateral and private-property tenure which have dominated most private bank credit scoring models. Thus, the certificate will rely on a combination of social as well as economic parameters accessible in real-time, i.e. the certificate will have a real-time score that indicates the score of the farmer at any particular time. The certification can be at the primary cooperatives and secondary cooperatives level. In addition to applying for finance, the certificate can be used to select candidates for management and leadership roles within the cooperative, in farming organisations as well as in government programs. For example, no persons can hold a leadership position without certification. These rules can, however, be enforced over a long period (a program can start from 5 years from now).

Establishment of Zimbabwe Good Agricultural Practices (ZiGAP)

Furthermore, I propose, through an act of parliament, the establishment of the Zimbabwe Good Agricultural Practices (ZiGAP) foundation/institution which would be responsible for individual, group, cooperative and collective certification in terms of meeting set safety and maintenance of standards in agricultural production. Learning from the Japanese GAP, I propose the government to head ZiGAP, work together with funding from the private sector and multilateral organisations. In addition to scoring farmers through the level of agricultural training received, many other self-evaluation variables should be collected (can be online or through a mobile application). ZiGAP can confer farmers with licences that prove they have attended and passed farmer training, ideological orientation, and they understand farming objectives.

It should also be responsible for preparation of Good Agricultural Practice (GAP) materials, dissemination of such materials, handling applications for the certificates, evaluation and recommendation for improvements of the farmer standards. ZiGAP should not be for cooperatives only but the whole national agricultural sector. Its standards should reflect international standards in order to make it easy for farmers and producers to penetrate international markets. The ZiGAP is discussed in greater detail elsewhere (see Muchetu et al., upcoming).

Chapter conclusion

The chapter presented lessons from Japan and contrasted them with the contemporary challenges in the Zimbabwe cooperative movement. This process helped to gain insights on the best ways to improve the sector which I presented through a perusal of the roles that the government, the community and the private sector had to play in cooperative development. The respective roles were formulated based on lessons learnt from Japan, modified and applied for Zimbabwe. I made some assumptions and highlighted essential prerequisites that would enable swift development and implementation of the alternative cooperative model which I presented.

The alternative cooperative model utilised the current structure and modified it, especially the gradual separation of the sectors such as agriculture, SACCOs and housing cooperatives to accommodate sector-specific considerations. I introduced a third type of cooperative that improves the cooperative appeal and improves membership and argued that the movement should start as a single-purpose cooperative, and then gradually grow into multi-purpose cooperatives. I propose interim financing through AgriBank, which should give way to CCB. Most importantly, I introduced the farmer and cooperative certification that can reduce transaction cost for banks, reduce information asymmetries (necessary for the CMS framework) and work as a motivation for farmers. Cooperative development hinges on the improvement of access to information, finance, education, training and skills development at local, provincial and national levels. The various suggestions provided in the new cooperative model aim to improve these issues.

The overall conclusion is that the market and the state cannot work alone. Neither can the state and the cooperative nor the community and the market as pairs in isolation. There need to be optimum or dynamic levels of interaction between the three. The state and the market have huge voices, and for so long, the community's voice could not be heard, exacerbating information asymmetries. Hence cooperatives have the highest potential of levelling the field for farmers. Based on the research question and the data analysis, as a set of lessons for the future development of agricultural cooperatives in Zimbabwe, I conclude that:

1. Cooperatives can survive within a capitalist production model if they are supported by robust state policies. This policy should ensure maximum community and minimum government involvement within the cooperative movement.

2. Land reforms are often radical and chaotic, resulting in temporary dysfunction of agricultural markets as well as the provision of public agricultural service delivery from the state and the markets. Thus, through self-help cooperatives, resettled farmers can access services otherwise unavailable.

3. From the data, two different categories of the cooperative movement can be identified, the old cooperative movement that entirely relies on the state (or donor) funding for survival on the one hand, and the new cooperative in the resettled areas which show a higher potential for growth if given a chance and support by the state on the other. The former more synonymous with the CA while the latter is more common in the resettled areas.

4. The state plays an essential role in the development of cooperatives through formulation and implementation of robust policies that would reduce exploitation from the markets, that give the movement autonomy and prioritises development ahead of political gains.

5. A new agricultural cooperative model for Zimbabwe should aim to establish a separate agricultural societies act administered in the Ministry of Agriculture. This new cooperative model places a focus on education, training and the need to develop farmer certification for short-run and long-run perspectives.

I firmly believe that more research needs to be carried out in terms of the financial contribution of the cooperative movement in Zimbabwe. This should start with the basic research of coming up with the cooperative registrar for all registered and active cooperatives in Zimbabwe. Once such a baseline survey of cooperatives has been achieved (preferably outside the government structures), then their contribution to, and their role in, alleviating poverty can be specified.

Overall summary of the book

The first four objectives of the book were leading to the attainment of the fifth objective that sought to restructure the cooperatives in pursuit of an alternative development path. The introductory Chapter One highlighted the nature of the problem in the agrarian areas and how it required rural people to unite and aggregate their voices when engaging the market. I argued that the nature and character of the AQ in the rural areas of Zimbabwe gave scope for the development of the cooperative model. Yet this has mostly been neglected by the government. The rationale for cooperatives is that:

- Cooperatives are legally easier to form as compared to other associative models such as investor-owned companies.
- The cooperative model provides a platform for dialogue between the central government, the farmers and the market. In such a platform, the state can provide subsidies, loans and tax exemptions while farmers can acquire information and negotiate prices in the market.
- The model tactfully eliminates the challenges brought by the middleman such that producer income becomes relatively higher while consumer prices become low.
- Cooperative models are generally more stable than private entities whose existence is affected by sudden market shocks.

Chapter Two described a series of procedures utilised in collecting data from Japan and Zimbabwe and how different theories supported these methods; mainly those put forward by Hayami Yujiro and Alexander Chayanov. Chapter Three described the trajectory of the JA development, which provided valuable insights into how the Japanese movement developed within the context of an expanding capitalist system and the role played by the government in this development. This process allowed us to derive lessons about the development phases of cooperatives, which Zimbabwe finds itself twenty years after its land reform. I described the development of the Zimbabwe cooperative movement from the pre-colonial to the current era in Chapter Four.

In Chapter Five, through data, the book mapped out the contemporary issues obtaining in the agricultural sector of Japan in order to further understand the challenges and opportunities that cooperatives face in the long term. Before I could recommend a new development path for Zimbabwe in Chapter Eight, I had to understand its contemporary cooperative movement, how they are structured, their challenges and opportunities. This process formed Chapters Six and Seven.

Bibliography

Akram-Lodhi, A. H., & Kay, C. (2010). Surveying the agrarian question (part 1): unearthing foundations, exploring diversity. *The Journal of Peasant Studies, 37*(1), 177–202. https://doi.org/10.1080/03066150903498838

Akwabi-Ameyaw, K. (1997). Producer cooperative resettlement projects in Zimbabwe: Lessons from a failed agricultural development strategy. *World Development, 25*(3), 437–456. https://doi.org/10.1016/s0305-750x(96)00106-4

Amin, S. (2012). Contemporary Imperialism and the Agrarian Question. *Agrarian South: Journal of Political Economy, 1*(1), 11–26. https://doi.org/10.1177/227797601200100102

Amin, S. (2018, May 3). *Interview with Samir Amin: "There is a Structural Crisis of Capitalism."* Portside: Materials of Interest to People on the Left. https://portside.org/2018-05-03/there-structural-crisis-capitalism

Ashkenazi, M., & Jacob, J. (2003). *Food Culture in Japan: Vol. Illustrated* (1st ed.). Greenwood Publishing Group.

Asuwa, S. (1962). Capital structure of agricultural cooperatives. *The Review of the Society of Agricultural Economics, 19*, 1–154. http://hdl.handle.net/2115/10805

Badiane, O. (1997). Agriculture, industrialisation, and food security in African countries by the year 2020. *Economic Transformation and Poverty Alleviation in African Countries into the 21st Century.*

Banaji, J. (1976). Chayanov, Kautsky, Lenin: Considerations Towards a Synthesis. *Economic and Political Weekly, 11*(40), 1594–1607. https://www.jstor.org/stable/4364979?seq=1#metadata_info_tab_contents

Banno, J. (1997). *The Political Economy of Japanese Society: The state of the market?* (illustrated). Oxford University Press.

Bernstein, H. (2004). "Changing Before Our Very Eyes": Agrarian Questions and the Politics of Land in Capitalism Today. *Journal of Agrarian Change, 4*(1–2), 190–225. https://doi.org/10.1111/j.1471-0366.2004.00078.x

Bernstein, H. (2005). Rural Land & Land Conflicts in Sub-Saharan Africa. In S. Moyo & P. Yeros (Eds.), *Reclaiming the land: The resurgence of rural movements in Africa, Asia and Latin America* (pp. 67–101). Zed Books.

Bernstein, H. (2009). V.I. Lenin and A.V. Chayanov: looking back, looking forward. *The Journal of Peasant Studies, 36*(1), 55–81. https://doi.org/10.1080/03066150902820289

Binswanger-Mkhize, H., & Moyo, S. (2012). Note II: Recovery and Growth of Zimbabwe Agriculture. In World Bank (Ed.), *Zimbabwe : From Economic Rebound to Sustained Growth : Growth Recovery*. World Bank.

Birchall, J. (1997). *The International Co-operative Movement.* Manchester University Press.

Birchall, J. (2005). Co-operative Principles Ten Years On. *Review of International Cooperation, 98*(2), 45–63.

Bond, P. (2008, December 10). *Lessons of Zimbabwe: An exchange between Patrick Bond and Mahmood Mamdani.* Links: International Journal of Socialist Renewal. http://links.org.au/node/815/9693

Bratton, M. (1987). The Comrades and the Countryside: The Politics of Agricultural Policy in Zimbabwe. *World Politics, 39*(2), 174–202. https://doi.org/10.2307/2010439

Buka. (2003). *The Buka report: A Preliminary Audit Report of Land Reform Programme.*

Byres, T. J. (1995). Political economy, the Agrarian Question and the Comparative Method. *Journal of Peasant Studies, 22*(4), 561–580. https://doi.org/10.1080/03066159508438589

Cathy. (2017, April 15). *How are nonprofits and co-ops different?* Cooperative Development Institute. https://cdi.coop/how-are-nonprofits-and-co-ops-different/

Chambati, W. (2017). Changing Forms of Wage Labour in Zimbabwe's New Agrarian Structure. *Agrarian South: Journal of Political Economy, 6*(1), 79–112. https://doi.org/10.1177/2277976017721346

Chayanov, A. V. (1925). *Organizatsiya krest'yanskogo khozyaistva, Reprinted in Cajanov (1967,1), translated in Chayanov (1966) as "The Peasant farm Organization."*

Chayanov, A. V. (1991). The Theory of Peasant Cooperations. In *Library: Second World series* (2nd ed.). Ohio State University Press.

Chayanov, A. V. (2018). What is the Agrarian Question (1917). *Russian Peasant Studies, 3*(2), 6–33. https://doi.org/10.22394/2500-1809-2018-3-2-6-33

Chitsike, F. (2003). *A Critical Analysis of the Land Reform Programme in Zimbabwe.* https://bit.ly/2F1sk3n

Chiweshe, M. K. (2011). *Farm Level Institutions in Emergent Communities in Post Fast Track Zimbabwe: Case of Mazowe District* [Rhodes University]. https://core.ac.uk/display/11984352

Collier, P., & Dercon, S. (2014). African Agriculture in 50 Years: Smallholders in a Rapidly Changing World? *World Development, 63*, 92–101. https://doi.org/10.1016/j.worlddev.2013.10.001

Cook, M. (2018). A Life-Cycle Explanation of Cooperative Longevity. *Sustainability, 10*(5), 1586. https://doi.org/10.3390/su10051586

Cook, M., Burress, M. J., & Buress, M. J. (2009). A cooperative lifecycle framework; a working paper. In *The Department of Agricultural and Applied Economics; University of Missouri* (Draft Papers). https://bit.ly/334qVBx

Cook, S., & Binford, L. (1986). Petty Commodity Production, Capital Accumulation, and Peasant Differentiation: Lenin vs. Chayanov in Rural Mexico. *Review of Radical Political Economics, 18*(4), 1–31. https://doi.org/10.1177/048661348601800401

Cornerstone. (2019, March 14). *3 Different Types of Management Styles in the Workplace.* Articles Resources. https://bit.ly/3cfLoHC

Cotterill, R. W. (1988). *Agricultural cooperatives: A unified theory of pricing, finance and investment.*

Davidow, J. (2019). *A Peace in Southern Africa: The Lancaster House Conference on Rhodesia, 1979.* Routledge. https://doi.org/10.4324/9780429045387

De Janvry, A. (1981). *The Agrarian Question and Reformism in Latin America.* Johns Hopkins University Press.

DGRV. (2009). *Germany cooperatives in Europe.*

Dogarawa, A. B. (2010). The Role of Cooperative Societies in Economic Development. *SSRN Electronic Journal.* https://doi.org/10.2139/ssrn.1622149

Dondo, A. M. (2012). *The cooperative model as an alternative strategy for rural development: a policy analysis case study of Kenya and Tanzania 1960-2009*. https://philpapers.org/rec/DONTCM-3

Dore, R. (2012). *Land Reform in Japan*. A&C Black.

Egan, D. (1990). Toward a Marxist Theory of Labor-Managed Firms: Breaking the Degeneration Thesis. *Review of Radical Political Economics, 22*(4), 67–86. https://doi.org/10.1177/048661349002200405

Eicher, C. K. (1995). Zimbabwe's maize-based Green Revolution: Preconditions for replication. *World Development, 23*(5), 805–818. https://doi.org/10.1016/0305-750x(95)93983-r

Emelianoff, I. V. (1948). *Economic theory of cooperation: Economic structure of cooperative organizations* (2nd ed., p. 271). Colombia University Press.

Esham, M., Kobayashi, H., Matsumura, I., & Alam, A. (2012). Japanese Agricultural Cooperatives at Crossroads: A Review. *American-Eurasian Journal of Agriculture & Environment Science of Agriculture & Environment Science, 12*(7), 954–959. https://doi.org/10.5829/idosi.aejaes.2012.12.07.1759

FAO. (1998). *Agricultural Cooperative Development: A Manual For Trainers*. https://doi.org/D/X0475E/2/9.01/500

FAO, & AUC. (2018). *Sustainable Agricultural Mechanization: A Framework for Africa*. Food and Agriculture Organisation. www.fao.org/publications

FAO, & IFAD. (2019, October 2). *Introducing the UN decade of Family Farming:* Food and Agriculture Organisation. http://www.fao.org/family-farming-decade/en/

FAOstat. (2016). *FAO Database*. FAO Meta-Database. http://www.fao.org/faostat/en/#data/QL

Francks, P. (1998). Agriculture and the state in industrial East Asia: the rise and fall of the food control system in Japan. *Japan Forum, 10*(1), 1–16. https://doi.org/10.1080/09555809808721600

George-Mulgan, A. (2005). Japan's Interventionist State: The Role of the MAFF. In *Japan's Interventionist State: The Role of the MAFF* (2nd ed., Vol. 2). Routledge Curzon: Taylor & Francis Group. https://doi.org/10.4324/9780203326640

George-Mulgan, A., & Honma, M. (2015). *The Political Economy of Japanese Trade Policy.* Palgrave Macmillan UK. https://doi.org/10.1057/9781137414564

Gluck, C. (1997, May 15). *Japan's Modern History: An Outline of the Period.* Asia Educators. http://afe.easia.columbia.edu/timelines/japan_modern_timelin e.htm

Godo, Y. (2014). The Japanese Agricultural Cooperative System : An Outline. *Agricultural Policy Platform (FFTC-AP)*, 1–5. https://ap.fftc.org.tw/article/678

Godo, Y. (2015). *The Treatment of Agricultural Cooperatives in the Antimonopoly Act in Japan.* https://ap.fftc.org.tw/article/979

Godo, Y. (2016). *Case Study of Unfair Trade Practices by a Japanese Agricultural Cooperative.* https://ap.fftc.org.tw/article/1000

GoJ. (1947a). Act on prohibition of private monopolization and maintenance of fair trade. In *The Constitution of Japan* (No. 54). Government of Japan. https://bit.ly/357KUl8

GoJ. (1947b). The constitution of Japan. In *Government of Japan (GoJ).* https://bit.ly/3gOxg8U

GoJ. (2018). *Statistical Handbook of Japan.* https://www.stat.go.jp/english/data/handbook/index.html

Govereh, J., Muchetu, R. G., Mvumi, B. M., & Chuma, T. (2019). Analysis of distribution systems for supply of synthetic grain protectants to maize smallholder farmers in Zimbabwe: Implications for hermetic grain storage bag distribution. *Journal of Stored Products Research, 84.* https://doi.org/10.1016/j.jspr.2019.101520

Grad, A. J. (1948). Land Reform in Japan. *Pacific Affairs, 21*(2), 115. https://doi.org/10.2307/2752510

Green, D. P. (2013, February 27). *Movement can give a strong voice to women | ICA.* International Cooperaitve Alliance. https://bit.ly/35gE25e

Gumede, W. (2018, October 15). *Lessons from Zimbabwe's failed land reforms.* University of Witwatersrand. https://bit.ly/2GDdmBf

Hairong, Y., & Yiyuan, C. (2013). Debating the rural cooperative movement in China, the past and the present. *Journal of Peasant Studies, 40*(6), 955–981.

https://doi.org/10.1080/03066150.2013.866555

Hanyani-Mlambo, B. T. (2000). Re-framing Zimbabwe's public agricultural extension services: institutional analysis and stakeholders views. *Agrekon, 39*(4), 665–672. https://doi.org/10.1080/03031853.2000.9523682

Hayami, Y., & Godo, Y. (2005). Income Distribution, Poverty, and Environmental Problems. In *Development Economics: From the Poverty to the Wealth of Nations* (pp. 191–239). Oxford University Press. https://doi.org/10.1093/0199272700.003.0008

Herbst, J. (1988). Societal Demands and Government Choices: Agricultural Producer Price Policy in Zimbabwe. *Comparative Politics, 20*(3), 265. https://doi.org/10.2307/421804

Hickey, S., & du Toit, A. (2007). Adverse Incorporation, Social Exclusion, and Chronic Poverty. In *Chronic Poverty Research Centre* (Vol. 81). Palgrave Macmillan. https://doi.org/10.1057/9781137316707.0012

Hirasawa, A. (2014). *Frame and Emerging Reform of Agricultural Policy in Japan.* https://ap.fftc.org.tw/article/691

Holyoake, G. J. (2016). *Self-Help by the people: The History of the Rochdale Pioneers.* London: Swan Sonnenschein and Co.; Routledge. https://doi.org/10.4324/9781315468853

Hoyt, A. (1996, February). *Cooperative Principles Updated.* Co-Operative Grocer Network. https://www.grocer.coop/articles/cooperative-principles-updated

ICA. (2019, May 17). *About International Co-operative Alliance | ICA.* International Cooperative Alliance. https://www.ica.coop/en/about-us/international-cooperative-alliance

Ichiya, T. (2016). Challenges of Agricultural Cooperatives in Japan for a Sustainable Future. In *FFTC Agricultural Policy Platform (FFTC-AP).* https://ap.fftc.org.tw/article/1114

Iliopoulos, C. (2017). The Evolution of Economic Theories of Agricultural Cooperatives : Schools of Thought , Theory Testing , and Systemic Implications. *Humboldt Workshop on the Study of Cooperatives, March.* https://doi.org/10.13140/RG.2.2.15265.20320

350

ILO. (2001). *Promotion of cooperatives: Fifth item on the agenda.*
https://bit.ly/2ESw73j

ILO. (2002). *R193 Promotion of Cooperatives Recommendation, 2002* (pp.
1–10).
https://www.ilo.org/dyn/normlex/en/f?p=NORMLEXPUB:1
2100:0::NO::P12100_ILO_CODE:R193

ILO. (2015, April 8). *Cooperatives and the Sustainable Development Goals:
A Contribution to the Post-2015 Development Debate. A policy brief.*
Cooperatives and the SDGs. https://bit.ly/2FHIRJV

Ishida, M. (2002a). Development of agricultural cooperatives in
Japan (1). *Bioresources Mie University, 28*(1), 19–34.

Ishida, M. (2002b). Development of agricultural cooperatives in
Japan (II) Revision of the Agricultural Cooperatives Law in
1996. *Bull. Fac. Bioresources, Mie University,* 35–47.

Ishida, M. (2002c). Development of agricultural cooperatives in
Japan (III) Historical perspective of the cooperative
development. *Bull. Fac. Bioresources, Mie University, 29*(1), 47–61.

Ishida, M. (2003). Development of Agriculture Cooperatives in
Japan (IV): Rural Communities and agricultural cooperatives at
the initial phase. *Bioresources Mie University, 29.*

Iwasaki, M. (1997). *Social History of agrarianism - life and the crossing of
the national polity.* Kyoto University Gakujutsu Shuppankai.

JA-GO. (2019, July 31). *Overview of JA Green Omi.* JA Green Omi.
http://www.jagreenohmi.jas.or.jp/about/index.html

JA-Group. (2012). *Enduring relations: Annual report 2012.*
https://bit.ly/3bvhYVn

Japan Times. (2016, July 30). *Japan's farming population falls below 2
million for first time: survey | The Japan Times.* Japan Times National
Outlook. https://bit.ly/3bzv1oY

Jayne, T. S., Haggblade, S., Minot, N., & Rashid, S. (2019).
Agricultural Commercialization, Rural Transformation and
Poverty Reduction: What have We Learned about How to
Achieve This? *Gates Open Research, 3,* 31.
https://doi.org/10.21955/GATESOPENRES.1115440.1

JCA. (2017). *The Japan Cooperative Alliance: What's new?* Japan
Cooperative Alliance (JCA). https://www.japan.coop/iyc2012/

Jefferson, R. A., Byth, D., Correa, C., Otero, G., & Qualset, C.

351

(1999). *Genetic use restriction technologies: Technical Assessment of the Set of New Technologies which Sterilize or Reduce the Agronomic Value of Second Generation Seed.* https://doi.org/10.5281/ZENODO.1477499

Johnson, D. (2009, February 12). *Mamdani, Moyo and 'deep thinkers' on Zimbabwe | Pambazuka News.* Pambazuka News: Voices of Freedom and Justice. https://bit.ly/33bnSan

Johnston, B. F., & Kilby, P. (1975). *Agriculture and Structural Transformation: Economic Strategies in Late-developing countries.* Oxford University Press.

Jossa, B. (2008). Gramsci and the Labor-Managed Firm. *Review of Radical Political Economics, 41*(1), 5–22. https://doi.org/10.1177/0486613408324421

Jossa, B. (2013). Alienation and the Self-Managed Firm System. *Review of Radical Political Economics, 46*(1), 5–14. https://doi.org/10.1177/0486613413488064

Jossa, B. (2014). Marx, Lenin and the Cooperative Movement. *Review of Political Economy, 26*(2), 282–302. https://doi.org/10.1080/09538259.2014.881649

Jossa, B. (2018). *A New Model of Socialism: Democratising Economic Production.* Edward Elgar Publishing. https://books.google.co.jp/books/about/A_New_Model_of_S ocialism.html?id=IyIWDwAAQBAJ&source=kp_book_descrip tion&redir_esc=y

Kako, T., Gemma, M., & Ito, S. (1997). Implications of the minimum access rice import on supply and demand balance of rice in Japan. *Agricultural Economics, 16*(3), 193–204. https://doi.org/10.1111/j.1574-0862.1997.tb00454.x

Kappes, A. (2014). *Cooperative development cooperation.*

Kawagoe, T. (1999). Agricultural Land Reform in Postwar Japan: Experiences and Issues. In *Policy Research Working Papers.* The World Bank. https://doi.org/10.1596/1813-9450-2111

Kinsey, B. H. (2004). Zimbabwe's Land Reform Program: Underinvestment in Post-Conflict Transformation. *World Development, 32*(10), 1669–1696. https://doi.org/10.1016/j.worlddev.2004.06.005

Klein, M. (2009). The cooperative works of Friedrich Wilhelm

Raiffeisen and its Christian roots. *Raiffeisen: DocPlayer*, 13.
https://docplayer.net/60241198-The-cooperative-work-of-
friedrich-wilhelm-raiffeisen-and-its-christian-roots.html

Kurimoto, A. (2004). Agricultural Cooperatives in Japan: An
Institutional Approach. *Journal of Rural Cooperation*, *32*(2), 111–
128. https://ideas.repec.org/a/ags/jlorco/59713.html

La Via Campesina. (2000, October 6). Bangalore Declaration Of
The Via Campesina. *3rd International Assembly*.
https://viacampesina.org/en/bangalore-declaration-of-the-via-
campesina/

Leavy, J., & Poulton, C. (2008). Commercialisations In Agriculture.
Ethiopian Journal of Economics, *16*(1).
https://doi.org/10.4314/eje.v16i1.39822

Lenin, V. I. (1921). *Lenin's Karl Marx: III: Marx's Economic Doctrine*.
The Marxist Archive.
https://www.marxists.org/archive/lenin/works/1914/granat/c
h03.htm

Lenin, V. I. (1923). *On Cooperation: Collected works* (Collected).
Progress Publishers.
https://www.marxists.org/archive/lenin/works/1923/jan/06.h
tm

Little, P. D., & Watts, M. (1994). *Living Under Contract: Contract
Farming and Agrarian Transformation in Sub-Saharan Africa*.
University of Wisconsin Press.
https://books.google.co.jp/books?id=uNPQrEdBXlgC

Lyimo, F. F. (2012). *Rural Cooperation: In the Cooperative Movement in
Tanzania*. Mkuki Na Nyota Publishers & African Books
Collective.

Mafeje, A. (2003). The agrarian question, access to land, and
peasant responses in Sub-Saharan Africa. In *United Nations
Research Institute for Social Development* (Issue 6).
https://bit.ly/31pCRy5

MAFF. (2015). *Report of survey on movement of agricultural structure*.
https://bit.ly/2F5rRgC

MAFF. (2017). *The 91st Statistical year book of agriculture, forestry and
fisheries*. https://bit.ly/328zxaP

MAFF. (2018). *2017 Summary of the Annual Report on Food, Agriculture and Rural Areas in Japan.* https://bit.ly/3gSkohX

MAFF. (2020). *Organizational structure of Ministry of Agriculture, Forestry and Fisheries.* https://bit.ly/3cE8x6H

Makamure, J., Jowa, J., & Mazuva, H. (2001). *Liiberalisation of agricultural markets* (Issue March, p. 73). s.n. http://www.saprin.org/zimbabwe/research/zim_agriculture.pdf

Makochekamwa, A. (2015). *An overview study of SACCO and cooperative movement in Zimbabwe.*

MAMID. (2012). *Comprehensive Agricultural Finance Framework (2012-2032).* http://extwprlegs1.fao.org/docs/pdf/zim149663.pdf

Mariani, M. (1906). *The cooperatives in social evolution (translated from Italian).*

Markowski, S., & Vanek, J. (1972). The General Theory of Labor-Managed Market Economies. *Economica, 39*(155), 339. https://doi.org/10.2307/2551860

Marx, K. (1973). On the first international. In *Inaugural address of the international working men's association.* McGraw-Hill. https://www.marxists.org/archive/marx/works/1864/10/27.htm

Marx, K. (1992). *Capital: A Critique of Political Economy* (Vol. 3). Penguin Classics.

Marx, K. (1996). Capital III: A critic of political economy. In F. Engels (Ed.), *Tim Delaney and M. Griffin in.* International Publishers.

Masiiwa, M. (2005). The Fast Track Resettlement Programme in Zimbabwe: Disparity between Policy Design and Implementation. *The Round Table, 94*(379), 217–224. https://doi.org/10.1080/00358530500082916

Masters, W. A. (1993). The scope and sequence of maize market reform in Zimbabwe. *Food Research Institute Studies, XXII*(3), 227–251. https://bit.ly/323mcAG

Matondi, P. (2012). *Zimbabwe's fast track land reform.* ZED Books. https://bit.ly/3lG3jvr

Mazoyer, M., & Roudart, L. (2006). *A History of World Agriculture: From the Neolithic Age to the Current Crisis.* NYU Press.

Mazwi, F., Chemura, A., Mudimu, G. T., & Chambati, W. (2019).

Political Economy of Command Agriculture in Zimbabwe: A State-led Contract Farming Model. *Agrarian South: Journal of Political Economy, 8*(1–2), 232–257. https://doi.org/10.1177/2277976019856742

Mazwi, F., & Muchetu, R. G. (2015). Out-grower sugarcane production post fast track land reform programme in Zimbabwe. *Ubuntu: Journal of Conflict Transformation, 4*(2), 17–48.

Mazwi, F., Muchetu, R. G., & Mudimu, G. T. (2021). Revisiting the trimodal agrarian structure as a social differentiation analysis framework in Zimbabwe: A Study. *Agrarian South: Journal of Political Economy, 10*(2).

McMichael, P. (2007). Peasant prospects in the neoliberal age. *New Political Economy, 11*(3), 407–418. https://doi.org/10.1080/13563460600841041

McMichael, P. (2013). Value-chain Agriculture and Debt Relations: contradictory outcomes. *Third World Quarterly, 34*(4), 671–690. https://doi.org/10.1080/01436597.2013.786290

McMichael, P. (2014). Historicizing food sovereignty. *The Journal of Peasant Studies, 41*(6), 933–957. https://doi.org/10.1080/03066150.2013.876999

Mhembwe, S., & Dube, E. (2017). The role of cooperatives in sustaining the livelihoods of rural communities: The case of rural cooperatives in Shurugwi District, Zimbabwe. *Jamba (Potchefstroom, South Africa), 9*(1), 341. https://doi.org/10.4102/jamba.v9i1.341

Mishima, T. (1992). Changes of the rice distribution and the functions of the food control system in Japan. *Agricultural Economics, 7*(1), 39–54. https://doi.org/10.1016/0169-5150(92)90020-y

Misininga. (1988). *Management and development functions of Zimbabwe agricultural marketing and supply cooperatives: A critical analysis of the problems and the strategies for improvement.*

Mkodzongi, G. (2016). Utilising 'African Potentials' to Resolve Conflicts in a Changing Agrarian Situation in Central Zimbabwe. In *What Colonialism Ignored* (pp. 75–102). Langaa RPCIG. https://doi.org/10.2307/j.ctvh9vtf6.8

Morishima, K. (2017). *Conspiracy to convert JA-Zenno.* Japan

Agricultural Communications.
https://www.jacom.or.jp/column/2017/
Morozumi, K. (1993). *The role of the agricultural cooperative credit system in Japanese agricultural finance.*
Moyo, S. (1992). *Land tenure issues in Zimbabwe during the 1990s.*
Moyo, S. / Centre for Applied Social Sciences Trust, UZ.
https://opendocs.ids.ac.uk/opendocs/handle/20.500.12413/10035
Moyo, S. (2000). The Political Economy of Land Acquisition and Redistribution in Zimbabwe, 1990-1999. *Journal of Southern African Studies, 26*(1), 5–28.
https://doi.org/10.1080/030570700108351
Moyo, S. (2004). The Land and Agrarian Question in Zimbabwe. In *Conference on 'The Agrarian Constraint and Poverty Reduction: Macroeconomic Lessons for Africa',Addis Ababa , 17-18 December, 2004* (Issue November). Semantic Scholar.
Moyo, S. (2005a). Land and Natural Resource Redistribution in Zimbabwe: Access, Equity and Conflict. *African and Asian Studies, 4*(1–2), 187–224.
https://doi.org/10.1163/1569209054547283
Moyo, S. (2005b). The land question and the peasantry in Southern Africa. In L. Maira (Ed.), *Politics and Social Movements in an Hegemonic World: Lessons from Africa* (pp. 146–163). CLACSO.
Moyo, S. (2007). Land in the Political Economy of African Development: Alternative Strategies for Reform. *Africa Development, 32*(4), 1–34.
https://doi.org/10.4314/ad.v32i4.57319
Moyo, S. (2011a). Changing agrarian relations after redistributive land reform in Zimbabwe. *Journal of Peasant Studies, 38*(5), 939–966. https://doi.org/10.1080/03066150.2011.634971
Moyo, S. (2011b). Land Concentration and Accumulation fter Redistributive Reform in Post-settler Zimbabwe. *Review of African Political Economy, 38*(128), 257–276.
https://doi.org/10.1080/03056244.2011.582763
Moyo, S. (2011c). Three decades of agrarian reform in Zimbabwe. *Journal of Peasant Studies, 38*(3), 493–531.
https://doi.org/10.1080/03066150.2011.583642

Moyo, S. (2015). Land Ownership Patterns and Income inequality in Southern Africa. *Pan Africa Conference on Inequalities in the Context of Structural Transformation 28-30 April 2014.* https://bit.ly/322D9dn

Moyo, S., Chambati, W., Murisa, T., Siziba, D., Dangwa, C., Mujeyi, K., & Nyoni, N. (2009). *Fast Track Land Reform Baseline Survey in Zimbabwe: Trends and Tendencies, 2005/06.*

Moyo, S., Chambati, W., & Siziba, S. (2014). *Agricultural subsidies policies in Zimbabwe: A review.*

Moyo, S., Jha, P., & Yeros, P. (2013). The Classical Agrarian Question: Myth, Reality and Relevance Today. *Agrarian South: Journal of Political Economy, 2*(1), 93–119. https://doi.org/10.1177/2277976013477224

Moyo, S., & Matondi, P. (2004). *Market-led land reform in Zimbabwe: Lessons and experiences.*

Moyo, S., & Nyoni, N. (2013). Changing Agrarian Relations after Redistributive Land Reform in Zimbabwe. In S. Moyo & W. Chambati (Eds.), *Land and Agrarian Reform in Zimbabwe* (pp. 195–250). CODESRIA. https://doi.org/10.2307/j.ctvk3gnsn.12

Moyo, S., & Skalness, T. (1990). Land Reform and Development Strategy in Zimbabwe: State Autonomy, Class and Agrarian Lobby. *Afrika Focus, 6*(3–4). https://doi.org/10.21825/af.v6i3-4.6125

Moyo, S., & Yeros, P. (2005a). Land occupations and land reform in Zimbabwe: Towards the national democratic revolution. In S. Moyo & P. Yeros (Eds.), *Reclaiming the Land: The Resurgence of Rural Movements in Africa, Asia and Latin America* (pp. 165–205). London.

Moyo, S., & Yeros, P. (2005b). *Reclaiming the Land: The Resurgence of Rural Movements in Africa, Asia and Latin America* (S. Moyo & P. Yeros (eds.)). Zed Books.

Moyo, S., & Yeros, P. (2007). The Radicalised State: Zimbabwe's Interrupted Revolution. *Review of African Political Economy, 34*(111), 103–121. https://doi.org/10.1080/03056240701340431

Moyo, S., & Yeros, P. (2013). The Zimbabwe Model: Radicalisation, reform and resistance. In S. Moyo & W. Chambati (Eds.), *Land*

357

and *Agrarian Reform in Zimbabwe: Beyond white settler capitalism* (pp. 331–358). CODESRIA & SMAIAS. https://doi.org/10.2307/j.ctvk3gnsn.15

Moyo, S., Yeros, P., & Jha, P. (2012). Imperialism and Primitive Accumulation: Notes on the New Scramble for Africa. *Agrarian South: Journal of Political Economy, 1*(2), 181–203. https://doi.org/10.1177/227797601200100203

Muchetu, R. G. (2018). Agricultural Land-Delivery Systems in Zimbabwe: A Review of Four Decades of Sam Moyo's Work on Agricultural Land Markets and their Constraints. *African Study Monographs, 57*, 65–94.

Muchetu, R. G. (2019a). Family farms and the markets: examining the level of market-oriented production 15 years after the Zimbabwe Fast Track Land Reform programme. *Review of African Political Economy, 46*(159), 33–54. https://doi.org/10.1080/03056244.2019.1609919

Muchetu, R. G. (2019b). Understanding Human Security in African Agrarian Societies: The Case for a Cooperative Model. *Journal of Human Security Studies, 8*(1), 20–43.

Mudege, N. N. (2005). *An ethnography of knowledge : knowledge production and dissemination in land resettlement areas in Zimbabwe: the case of Mupfurudzi* [Wageningen University]. https://bit.ly/3272LXA

Muir-Leresche, K., & Muchopa, C. (2006). Agricultural marketing. In M. Rukini, P. Tawonezvi, & C. Eicher (Eds.), *Zimbabwe's agricultural revolution revisited* (Vol. 2, pp. 299–317). University of Zimbabwe Publishers.

Muir, K., & Takavarasha, T. (1989). Pan-territorial and pan-seasonal pricing for maize in Zimbabwe. In G. Mudimu & R. H. Bernstein (Eds.), *Household and national food security in Southern Africa* (pp. 103–124). University of Zimbabwe (UZ) Publications/ Michigan State University (MSU). https://bit.ly/3jXEf11

Mujeyi, K. (2010). *Livelihoods after Land Reform in Zimbabwe Working Paper 1 Emerging Agricultural Markets and Marketing Channels within Newly Resettled Areas of Zimbabwe* (No. 1). https://bit.ly/2Yh2Vte

Murisa, T. (2009). *An Analysis of Emerging Forms of Social Organisation*

and Agency in the Aftermath of "fast track " Land Reform in Zimbabwe [Rhodes University].
https://core.ac.uk/download/pdf/145052251.pdf

MYDEC. (2005). *The revised government policy on cooperative development.*

Nagatani, T. (2015). Succession of Farmlands to Non-Family Successors: Options for the Young Generation of Farmers. *FFTC Agricultural Policy Platform (FFTC-AP).* https://ap.fftc.org.tw/article/874

Nakane, C. (1967). *Kinship and Economic Organization in Rural Japan* (Reprint). Berg Publishers.

Nakane, C. (1970). *Japanese Society* (illustrate). University of California.

Nishikawa, K. (2014). *Japanese Agricultural Policy Reforms after FY 2014 (Part 1)* . https://ap.fftc.org.tw/article/681

Nishikawa, K. (2015). Amendments of the Agricultural Cooperative Act in 2015. *FFTC Agricultural Policy Platform (FFTC-AP).* https://ap.fftc.org.tw/article/1037

Norinchukin. (2019). *JA Cooperative Bank Report 2019.* https://bit.ly/3bxaz8b

Nyafwa, K. (2019). *Credit scoring in Kenya.* Doshisha University.

Oczkowski, E., Krivokapic-Skoko, B., & Plummer, K. (2013). The meaning, importance and practice of the co-operative principles: Qualitative evidence from the Australian co-operative sector. *Journal of Co-Operative Organization and Management, 1*(2), 54–63. https://doi.org/10.1016/j.jcom.2013.10.006

OECD. (2009). Evaluation of agricultural policy reforms in Japan. In *Evaluation of Agricultural Policy Reforms in Japan* (Vol. 9789264061). https://doi.org/10.1787/9789264061545-en

Ogura, T. (1966). RECENT AGRARIAN PROBLEMS IN JAPAN. *The Developing Economies, 4*(2), 151–170. https://doi.org/10.1111/j.1746-1049.1966.tb00777.x

Ortmann, G. F., & King, R. P. (2007). Agricultural Cooperatives I: History, Theory and Problems. *Agrekon, 46*(1), 18–46. https://doi.org/10.1080/03031853.2007.9523760

Otsuka, K., & Kalirajan, K. (2010). Community, market and state in development. In *Community, Market and State in Development.*

Palgrave McMillan. https://doi.org/10.1057/9780230295018

Patnaik, U. (1988). Ascertaining the economic characteristics of peasant classes-in-themselves in rural India: A methodological and empirical exercise. *The Journal of Peasant Studies, 15*(3), 301–333. https://doi.org/10.1080/03066158808438365

Patnaik, U., Moyo, S., & Shivji, I. G. (2011). *The Agrarian Question in the Neoliberal Era: Primitive Accumulation and the Peasantry* (Illustrate). Pambazuka Press.

Pinto, A. (2009). *Agricultural cooperatives and farmer's organizations: Role in rural development and poverty reduction.* https://bit.ly/3lc4m5Z

Poulton, C., Davies, R., Matshe, I., & Urey, I. (2002). *A review of Zimbabwe's agricultural economic policies: 1980-2000* (No. 12; 01). https://bit.ly/3ixIpvO

Prakash, D. (2003). Development of agricultural cooperatives: Relevance of Japanese experiences to developing countries. *Strengthening Management of Agricultural Cooperatives in Asia IDACA-Japn,* 17.

Prinz, M. (2002, July). German Rural Cooperatives, Friedrich-Wilhelm Raiffeisen and the Organization of Trust. *Agricultural, Cattle Breeding and Fishing Cooperativism and Associationism in Europe and Latin America.* https://bit.ly/2YW0Rae

Romdhane, M. Ben, & Moyo, S. (2002). Peasant organizations and rural civil society in Africa: An Introduction. In M. Ben Romdhane & S. Moyo (Eds.), *Peasant Organisations and the Democratisation Process in Africa* (1st ed., Vol. 2, p. 1). CODESRIA.

Royer, J. (1999). Cooperative Organizational Strategies: A Neo-Institutional Digest. *Journal of Cooperatives, 14,* 1–24. https://doi.org/10.22004/ag.econ.46367

Royer, J. (2014). The Theory of Agricultural Cooperatives: A Neoclassical Primer. *Faculty Publications: Agricultural Economics, 23*(1), 1–56. https://digitalcommons.unl.edu/ageconfacpub/123

Rusike, J., & Sukume, C. (2006). Agricultural input supply. In M. Rukuni, P. Tawonezvi, & C. Eicher (Eds.), *Zimbabwe's agricultural revolution revisited* (2nd ed., Vol. 2, pp. 279–295). University of Zimbabwe (UZ) Publications.

https://opendocs.ids.ac.uk/opendocs/handle/20.500.12413/10 090

Sachikonye, L. M. (1995). Civil society, social movements and Democracy in Southern Africa. *Innovation: The European Journal of Social Science Research*, *8*(4), 399–411. https://doi.org/10.1080/13511610.1995.9968465

Sachikonye, L. M. (2005). The Land is the economy. *African Security Review*, *14*(3), 31–44. https://doi.org/10.1080/10246029.2005.9627368

Sadomba, Z. W. (2011). *War Veterans in Zimbabwe's Revolution: Challenging Neo-colonialism & Settler* (illustrate). Boydell & Brewer Ltd, 2011.

Saito, Y. (2018). *Characteristics of agricultural cooperatives' self-reform and its challenges : with a special focus on primary cooperatives' efforts to enhance agricultural production.* https://www.nochuri.co.jp/

Samkange, S. J. T. (1980). *Hunhuism or ubuntuism: A Zimbabwe indigenous political philosophy.* Graham Pub.

Samuel, M. (2018). Mobile Phone Use by Zimbabwean Smallholder Farmers: A Baseline Study. *The African Journal of Information and Communication*, *22*, 29–52. https://doi.org/10.23962/10539/26171

Samuelson, P. (2012, February 14). *Today's status of the New Institutional Economics and its differences to the Neo-Classical Economics.* New Institutional Economics at University of Bifrost. https://bit.ly/2ETz8zQ

Schaars, M. A. (1971). *Cooperatives, Principles and Practices* (2nd ed., Vol. 1). The University of Wisconsin.

Schwettmann, J. (2000). *Cooperatives and Employment in Africa.* https://bit.ly/3h3nbEY

Scoones, I., Marongwe, B., Mavedzenge, F., Murimbarimba, F., & Sukume, C. (2010). *Zimbabwe's Land Reform: Myths & Realities* (I. Scoones (ed.)). James Currey.

Scott, J. C. (1985). *Weapons of the Weak: Everyday Forms of Peasant Resistance* (Illustrate, Vol. 12). Yale University Press.

Scott, J. C. (1998). *Seeing Like a State: How Certain Schemes to Improve the Human Condition Have Failed.* Yale University Press.

361

Scott, J. C. (2012). Two Cheers for Anarchism. In *Two Cheers for Anarchism*. Princeton University Press. https://doi.org/10.1515/9781400844623

Shanin, T. (1981). Marx, Marxism and the Agrarian Question: I Marx and the Peasant Commune. *History Workshop Journal, 12*(1), 108–128. https://doi.org/10.1093/hwj/12.1.108

Shanin, T. (2009). Chayanov's treble death and tenuous resurrection: an essay about understanding, about roots of plausibility and about rural Russia. *The Journal of Peasant Studies, 36*(1), 83–101. https://doi.org/10.1080/03066150902820420

Shivji, I. G. (2009). Accumulation in an African periphery: A theoretical framework. In *Accumulation in an African Periphery: A Theoretical Framework*. Mkuki na Nyota Publishers.

SMAIAS. (2015). *Inter-district household survey: Follow up to the baseline survey*.

SMECD. (2017). *SMECD: About us*. Ministry of Small and Medium Entreprises and Cooperative Development. http://www.smecd.gov.zw/index.php/about-smes

Stack, J. L. (1994). The distributional consequences of the smallholder maize revolution. In M. Rukuni & C. K. Eicher (Eds.), *Zimbabwe's Agricultural Revolution* (Issue 4, pp. 258–269). University of Zimbabwe Publishers.

Stack, J. L., & Sukume, C. (2006). Rural poverty: challenges and opportunities. In M. Rukuni, P. Tawonezvi, & C. Eicher (Eds.), *Zimbabwe's agricultural revolution revisited* (2nd ed., Vol. 2, pp. 557–570). University of Zimbabwe (UZ) Publications.

Stiglitz, J. E. (2002). Information and the Change in the Paradigm in Economics. *American Economic Review, 92*(3), 460–501. https://doi.org/10.1257/00028280260136363

Teruoka, S. (2008). *Agriculture in the Modernization of Japan, 1850-2000*. Manohar Publishers & Distributors.

Thomson, D. J. (1994). *Weavers of Dreams: Founders of the Modern Co-operative Movement* (Vol. 2, p. 152). Regents University of California Press.

Thorner, D. (1965). A post-Marxian theory of peasant economy: The school of A V Chayanov. *Economic and Political Weekly, 17*(5-6–7), 227–234.

Thorner, D., Kerblay, B., Smith, R. E. ., & Shanin, T. (1986). *A.V. Chayanov on the Theory of Peasant Economy* (Illustrate). University of Wisconsin Press.

Timmer, P. C. (2009). *A World Without Agriculture: The Structural Transformation in Historical perspective.* American Enterprise Institute (AEI) Press.

Toendepi, S. (2018). *Reconfigured Agrarian Relations in Zimbabwe.* Langaa RPCIG.

USDA. (2011). *The structure of cooperatives: Cooperative information report.* https://www.rd.usda.gov/files/CIR45_3.pdf

Utete, C. M. B. (2003). *Report of the presidential land review committee and presidential land review committee.* https://bit.ly/34XCBZj

Vincent, V., & Thomas, R. G. (1961). *An agricultural survey of Southern Rhodesia: Part I: agro-ecological survey.*

Vitoria, B., Mudimu, G., & Moyo, T. (2012). Status of Agricultural and Rural Finance in Zimbabwe. *FinMark Trust, July,* 95. https://bit.ly/3jWX6tb

Wedig, K., & Wiegratz, J. (2018). Neoliberalism and the revival of agricultural cooperatives: The case of the coffee sector in Uganda. *Journal of Agrarian Change, 18*(2), 348–369. https://doi.org/10.1111/joac.12221

Weiner, D. (1988). Agricultural transformation in Zimbabwe: Lessons for South Africa after apartheid. *Geoforum, 19*(4), 479–496. https://doi.org/10.1016/s0016-7185(88)80019-8

Wiggins, S., Argwings-Kodhek, G., Leavy, J., & Poulton, C. (2011). *Small farm commercialisation in Africa: Reviewing the issues.* https://bit.ly/327mhlB

Winn, J. (2013, June 25). *The association of free and equal producers.* Joss Winn Organisation. https://bit.ly/3io1266

World Bank. (1989). *Zimbabwe agricultural sector review.*

World Bank. (2017, June 21). *World Bank Metadata.* World Bank Metadata. https://data.worldbank.org/indicator/AG.LND.AGRI.ZS?locations=ZW

World Bank. (2019a). *Zimbabwe Public Expenditure Review with a Focus on Agriculture.* https://bit.ly/2Fbaefe

World Bank. (2019b, October 12). *Rural population (% of total population) - Sub-Saharan Africa | Data*. World Bank Metadata. https://bit.ly/32jxlgX

Yamashita, K. (2009). The Agricultural Cooperatives and Farming Reform in Japan (1) . *The Tokyo Foundation for Policy Research*. https://www.tkfd.or.jp/en/research/detail.php?id=79

Yamashita, K. (2015a). The Political Economy of Japanese Agricultural Trade Negotiations. In A. George-Mulgan & M. Honma (Eds.), *The Political Economy of Japanese Trade Policy* (1st ed., Vol. 1, pp. 71–93). Palgrave Macmillan UK. https://doi.org/10.1057/9781137414564_3

Yamashita, K. (2015b, April 20). *A First Step Toward Reform of Japan's Agricultural Cooperative System*. Nippon.Com. https://www.nippon.com/en/currents/d00169/

ZimStat. (2012). *Zimbabwe national population census (2012)*. http://www.zimstat.co.zw/wp-content/uploads/publications/Population/population/census-2012-national-report.pdf

ZimStats. (2018). *Poverty Analysis*.

Zumbika, N. (2006). Agricultural finance: 1990-2004. In M. Rukuni, P. Tawonezvi, & C. Eicher (Eds.), *Zimbabwe's agricultural revolution revisited* (2nd ed., Vol. 2, pp. 339–353). University of Zimbabwe (UZ) Publications. https://bit.ly/32jxAZp

Index

A

accountability, 64, 137, 218, 271, 282, 336
accumulation, 18, 41, 50, 83, 199, 229, 238, 297, 298, 300, 301, 302, 303, 310, 327
acquisition, 1, 79, 120, 123, 185
    land acquisition, 117, 119, 120, 121
adherence, 37, 140, 165, 166, 335, 336
ageing, 105, 160, 196, 197, 199, 200, 241
agency, 4, 23, 43, 45, 47, 81, 85, 118, 119, 193, 203, 228, 318
    farmer agency, 75, 193, 322
agrarian question, 4, 5, 6, 8, 9, 24, 25, 53, 267, 311, 321
    AQ, 4, 5, 6, 7, 8, 9, 12, 15, 18, 23, 24, 26, 53, 67, 265, 275, 282, 305, 343
AgriBank, 142, 331, 337, 338, 342
agribusiness, 144, 178, 199
agricultural basic, 326
Agricultural Basic Law, 87, 91, 92, 135
agricultural commodities, 243
agricultural education, 321
    education and training, 138, 261, 322, 323, 335, 337, 339
    farmer education, 335
    skills, 207, 295, 296, 297, 321
Agricultural Finance Cooperation, 118
    AFC, 118, 141, 142
Agricultural Societies Act, 343
amend, 318
    amendments, 73, 93, 94, 122, 135, 191, 195, 206
apolitical, 190, 230, 324
Asia, 7, 12, 13, 54
    Asian, 11, 56, 70, 285
    Asian Green Revolution, 12
autocratic management, 182, 185
autonomy, 38, 39, 53, 100, 103, 173, 318, 320, 343

B

bad debtors, 33, 283, 284, 335
beneficiaries, 35, 122, 123, 124, 125, 144, 166, 219, 321

Brazil, 213, 224, 246, 247
British-Indian, 84, 109, 114, 144, 145, 248, 327
by-laws, 114, 138, 223, 225, 246
    cooperative by-laws, 136, 225, 270, 271, 272

C

capacity building, 25, 84, 104, 232, 310
Central Association of Cooperative Unions, 336
Central Cooperative Bank, 101, 337
    CCB, 101, 105, 337, 338, 342
challenges faced, 155, 243
    constraints, 97, 149, 156, 199, 241, 242
    impediments, 53, 156, 216, 229, 306
    malpractices, 218
Chiba, 19, 21, 22, 154, 161, 162, 202, 294
classification, 62, 251, 258
collateral, 242, 281, 282, 283, 293, 337, 340
colonial, 113
    colonial era, 19, 110, 113, 129
command agriculture, 134, 230, 287
Commercial Farmer's Union, 130
    CFU, 130, 132
communication, 36, 80, 144, 236, 323, 330
communist, 16, 81, 101, 158, 170, 178
communism, 16, 53, 79, 158
community development, 12, 67, 71, 194
Community-Market-State, 4, 26, 311, 317
    CMS, 4, 27, 28, 29, 174, 177, 266, 275, 318, 321, 325, 342
consumer cooperative, 32, 59, 160, 161, 166, 168, 173, 174, 182, 186
cooperative act, 221, 248
cooperative banking, 34, 98, 181, 186
cooperative solidarity, 52, 65, 100, 108, 193
Cooperatives Societies Act, 135
    CSA, 135, 136, 217, 227, 230, 231, 234

365

D

debt collection, 271, 283
decentralisation, 236
degenerate, 108, 155, 191, 195, 322
    degeneration, 46, 50, 51, 52, 65, 66,
        108, 191, 197, 314
democratic member control, 38, 113
deregulation, 97, 131, 195
disguised workers, 56
distribution of wealth, 31
dividends, 25, 109, 149, 191, 194, 244,
    328
donor syndrome, 212, 316, 320

E

Ecocash, 293, 318
economic democracy, 66
Economic Structural Adjustment
    Programmes, 2
    economic adjustment programs, 114
    ESAP, 2, 3, 7, 28, 29, 91, 112, 118,
        119, 120, 126, 127, 128, 129, 131,
        132, 133, 135, 141, 142, 145, 235,
        315
equality
    inequality, 7, 26
    less equality, 100, 109
    more equality, 100, 109, 327
Equality, 36
Establishment of collectives, 116
    collectivisation, 27, 33, 238
exploitation, 58
    capitalist exploitation, 6, 11, 61
    economic exploitation, 58
    exploitative, 56, 82

F

farmer attitude, 212, 244, 316, 323
farmer certificate, 83, 334
    certificate, 137, 334, 340
    certified farmer, 192, 334
    certified farmers, 166, 191
    farmer certification, 136, 334, 340,
        343
farmer certification
    cooperative certification, 332, 333,

334, 338, 342
Fast-Track Land Reform Programme, 3
    FTLRP, 121, 123
feedback, 163, 269
feudal system, 56, 76
food agency, 95
food control system, 95, 129, 130
food sovereignty, 4, 9, 135

G

gender, 259
globalisation, 39, 40, 170, 232
GMO, 134, 162
Good Agricultural Practices, 331, 341
Grain Marketing Board, 129

H

heterogeneity, 43, 94, 106, 156
horizon, 46, 109
horizontal integration, 33, 56
housing cooperatives, 34, 146, 220, 223,
    229, 230, 231, 342
Hurungwe, 1, 2, 3, 111
Hyogo, 12, 19, 21, 160
hyperinflationary, 82

I

imperialism, 16, 77
import substitution, 2, 7
industrialisation, 6, 24, 86
inefficiency, 31, 262, 267
information asymmetries, 12, 27, 28, 29,
    43, 174, 175, 177, 239, 276, 316, 318,
    322, 342
irrigation facilities, 20, 86, 93

J

Jiyu, 90
Juru village, 19, 21

366

K

Keizairen, 88
kinship, 29, 113, 221
Kokkai, 87
Kumiaika, 95

L

labour-managed firms, 48, 50
Lancaster House Conference, 117
  LHC, 117, 118
land administration system, 241
Land Apportionment Act, 113, 119
land redistribution, 3, 10, 14, 118, 119, 124
land tenure insecurity, 241
landlord, 76

M

MAFF, 94
market liberalisation, 7, 277
maSvikiro, 2
methodology, 20
  PCA, 253, 254
mistrust, 149, 248, 266, 308
Mugabe, 3, 229, 230, 324
multipurpose, 101, 314, 338

N

Natural Region, 126
new cooperative model, 5, 19, 155, 313, 314, 333, 342, 343
new dispensation, 229, 324
New Zimbabwe Cooperative, 339
Norinchukin Bank, 99

P

partnership, 85, 168, 181
peasant differentiation, 54, 55, 61
political manipulation, 212, 224, 226

portfolio problems, 109
poverty reduction, 35, 122
progressive, 25, 52, 72, 141, 179, 233

R

radicalism, 70, 121
Raiffeisen movement, 33
registration of cooperatives, 216
resettlement, 116, 122, 125, 238, 321
restructure, 5, 34, 46, 102, 151, 343
Rochdale, 15, 32, 33, 34, 109

S

Sam Moyo, 14
Sanbu village, 162
sanctions, 114
Savings and Credit Cooperative Organisations, 24
  SACCO, 24, 144, 186, 231, 321
Schulze-Delitzsch, 15, 34
self-exploitation, 50
share capital, 73, 100, 231
Shiga, 19, 21, 160, 165
small-holder, 13, 132, 214
solutions, 9, 17, 34, 39, 46, 65, 210, 213, 219, 223, 239, 241, 247
squatting, 119, 120, 121
state subsidies, 320
storage facilities, 73, 302, 308
survey participants, 20, 44, 132, 153, 154, 206, 207, 220, 264, 271
sustainability, 24, 40, 97, 108, 124, 125, 134, 137, 149, 151, 169, 203, 209, 213, 232, 272, 299, 310, 318, 337

T

Theory of Peasant Cooperative, 23
  Chayanovian, 15, 33, 41, 60, 62, 65, 67, 169, 251, 257, 289, 317
  neoclassical, 30, 41, 43, 64, 65
  NIE theory, 17, 65
transparency, 133, 218, 266, 308

U

UDI, 126
Ujamaa, 27
unemployment, 32, 149
unregistered cooperatives, 216, 315

V

vertical integration, 42, 60

W

War Veterans Associations, 3
welfare, 26, 36, 40, 41, 50, 93, 94, 106,
    165, 170
WMF. *See* labour-managed firms

X

Xanadu, 19, 21, 207, 219, 230, 237, 238,
    239, 244, 247, 295, 329

Z

Zenchu, 85
Zenkoren, 102
ZiGAP, 331, 332, 338, 341
ZIMACE, 133
Zim-Japan similarities, 69, 70, 210, 311,
    325, 326

368

Printed in the United States
by Baker & Taylor Publisher Services